A STUDY OF THE CAT————

with reference to human beings

third edition

Warren F. Walker, Jr., Ph.D.

Professor of Biology, Oberlin College,
Oberlin Ohio

1977
W. B. SAUNDERS COMPANY
Philadelphia, London, Toronto

W. B. Saunders Company: West Washington Square
Philadelphia, PA 19105

1 St. Anne's Road
Eastbourne, East Sussex BN21 3UN, England

1 Goldthorne Avenue
Toronto, Ontario, M8Z 5T9, Canada

Permission to use the illustrations shown on the cover has been granted by Dr. R. F. Sis and Dr. John Hope.

A Study of the Cat
with Reference to Human Beings

ISBN 0-7216-9093-9

Last digit is the print number: 9   8   7   6   5   4   3   2   1

PREFACE

This laboratory manual is designed for the many students who wish a thorough understanding of mammalian anatomy. It will be of use in certain comparative anatomy courses where the laboratory emphasis is on mammalian anatomy, and in prenursing and other courses where there is a need to understand human structure but it is not feasible to dissect a human cadaver. Cats are favorable material because they are large enough to dissect easily, their organs are in the adult stage of development, and, on balance, their anatomy is closer to the condition in human beings than is that of other readily available animals. The descriptions of the eye, brain, and heart are based on sheep material, for this material is easy to get and easier to study than comparable cat organs. Comparisons have been made between the cat and human beings so that the student will be aware of major differences. It can be assumed that our anatomy substantially resembles a cat's if differences are not mentioned.

Mammals are the product of a long evoutionary history, so some knowledge of their phylogeny is helpful to understand their present structure. Brief evolutionary summaries are given so that the student who does not have time for a full comparative anatomy course can gain some appreciation for the origin of mammalian structures. Brief discussions of the functional significance of certain structures and anatomical interrelationships are also included to give the student yet another level of understanding of mammalian form and function.

I have taken the opportunity in this new edition to make many small changes throughout the book, and major revisions in certain areas. The discussion of the skull includes more information on the mechanical forces involved in its evolution. Several groups of muscles have been redescribed to make their dissection easier. An expanded discussion of the biomechanics of the jaw will help students understand differences in the jaw and the jaw musculature of a carnivore (cat), a gnawing herbivore (rabbit), and an omnivore (human being). The discussion of mammalian circulation has been rewritten to include a comparison of fetal, neonatal, and adult circulation. This will enable students to understand the significance of fetal remnants that they will see during the adult dissection. New information is presented on the functional significance of the cat's carotid rete mirabile. Comparisons with human structures have been expanded.

The section on anatomical terminology has been moved from the end to the beginning of the book, where students are more likely to see and use it at the outset of the course. Favored terms are set in boldface italic type when first used. For the most part, these are terms agreed upon in the most recent edition of the *Nomina Anatomica Veterinaria* (1973). Veterinary terms, in turn, are based upon those used in human anatomy (*Nomina Anatomica Parisiensia*, 1955), differing with respect to some terms for direction (e.g., human "superior" = quadruped "cranial"), and in the few cases in which human and quadruped structures differ (e.g., the human pectoralis major = the cat pectoralis superficialis, for this muscle is not the larger of the pectorals in quadrupeds). Sometimes a well-established English term has been favored over the *Nomina Anatomica* term, but in such cases the *Nomina* term is cited afterward in light face italics; e.g., *kidney (ren)*. *Ren* is not used as a noun in English but forms the basis for the adjective renal.

An appendix has been added on Word Roots, in which the more important Latin and Greek words used in building anatomical terms are given and defined and examples of their use cited. To many students, anatomy approaches a foreign language. It is my hope that this list will take some of the mystique out of anatomical terminology and make the terms easier, and perhaps even fun, to learn.

Many figures are included so that the manual can also serve as a partial atlas. Figures can save a considerable amount of the student's time, both in finding the structures and in serving as a record of his observations. But the student should be cautioned that figures should not serve as a substitute for careful dissection and observation. Most instructors give several examinations on the specimens, which helps to insure that careful work is done. Many of the figures show not only the organs being considered but the relationship of these organs to surrounding parts. This should facilitate finding and remembering the structures concerned.

This new edition has given me the opportunity to make many improvements in the illustrations. Thirty-four new figures are included, 13 of which replace previous figures and 21 of which are new additions. The total number of figures is now 114. Minor changes have been made in many other figures. Most of the drawings of the circulatory system show both the arteries and veins together, for this is the way they are seen by the student. The favorable reception of earlier editions has made it possible to use two colors to distinguish arteries and veins.

I am indebted to many for help in preparing the third edition of this book. Students and colleagues have sent comments and useful suggestions to me or to the publishers. I wish to thank all of them, known and unknown to me, and I earnestly hope that they will continue to call my attention to any errors or deficiencies that they may find in this edition. Their past support and encouragement have greatly increased the merit of this manual. I continue to be indebted to Dr. Douglas B. Webster of the Louisiana State University Medical Center for the section on the Dissection of the Brain that he prepared for an earlier edition. He has succeeded admirably in integrating structure and function and in providing an excellent introduction to neuroanatomy. Mr. J. Paul Nail, a student in anatomical art at Oberlin College, has prepared the new illustrations. His artistry and professionalism have raised the quality of the figures in this book, and I am very grateful to him. My wife has again been most helpful in lending encouragement and in proofreading the materials in their various stages of writing. I also wish to thank the staff of the W. B. Saunders Company, who have been most helpful and encouraging in preparing this book.

WARREN F. WALKER, JR.

CONTENTS

Chapter Five

THE SENSE ORGANS 82

Chapter Six

THE NERVOUS SYSTEM 93

A NOTE TO THE STUDENT:
Anatomical Terminology

Courses in Anatomy are sometimes compared to a foreign language in that a large vocabulary of often unfamiliar words is introduced. This cannot be avoided if one is to describe and discuss the parts of an animal and their locations. Most of the terms are based on Latin or Greek roots, and as you become familiar with the more common roots you will recognize the terminology for the shorthand that it is. The root "chondro," for example, always means cartilage, and it is used in many combinations: Chondrichthyes (cartilaginous fish), chondrocranium (cartilaginous braincase), perichondrium (connective tissue around cartilage), chondrocyte (cell in cartilage). There are also certain conventions, described below, that make the terminology more rational.

TERMS FOR ORGANS

As anatomists described the structure of animals, they used terms for organs that suited their fancy. Commonly an organ was described by its appearance in a human being, for our anatomy was one of the first to be studied. Over the centuries a plethora of names has been proposed, and many organs have a long list of synonyms. In order to bring some order to the developing chaos, human anatomists throughout the world have agreed upon codes of terminology, the most recent being the *Nomina Anatomica Parisiensia* (NAP, 1955). Terms are in Latin or Greek, but they are often translated into the vernacular of each language. Veterinary anatomists have agreed upon a *Nomina Anatomica Veterinaria* (most recently in 1973) in which they have brought most of the terminology for other mammals into agreement with human terminology, the major exception being certain terms for direction which are naturally different between a biped and quadruped (see below). No conventions have been agreed upon for nonmammalian vertebrates, but the tendency is to apply mammalian terms whenever applicable.

In this edition of *A Study of the Cat*, I have used *Nomina Anatomica Veterinaria* terms for cat structures in most cases. Favored terms are placed in boldface italics when first used. In the few cases for which this is not a *Nomina Anatomica* term, the official term is given in light face italics, e.g., *kidney (ren)*. The official term in this case is seldom used in English as a noun, but it forms the basis for the common adjective, renal. I have introduced synonyms only when they are in common use, in which case the synonym is given in regular type, e.g., *cleidobrachialis muscle* (clavodeltoid), or when the homologous human structure has a different name, e.g., *cranial vena cava* (superior vena cava). Terms based on the names of individuals are avoided in official terminology, but some in common use have been included, e.g., *auditory tube* (eustachian tube).

TERMS FOR DIRECTIONS, PLANES, AND SECTIONS

It is necessary when discussing the anatomy of animals to have terms that can be used to describe the location of parts. The following are the ones recommended in the *Nomina Anatomica Veterinaria* for use in a quadruped. Most of them are illustrated in the accompanying diagram.

Terms for Direction

Many terms for direction are the same in comparative and human anatomy, but there are certain differences occasioned by our upright posture. A structure toward the head end of a quadruped is described as *cranial* or *rostral* (e.g., cranial vena cava); one toward the tail, as *caudal.* Comparable positions in human beings are described as *superior* (e.g., superior vena cava) and *inferior* (see Figure N–1).

A structure toward the back of a quadruped is *dorsal;* one toward the belly is *ventral.* The terms dorsal and ventral are not used in human anatomy; comparable

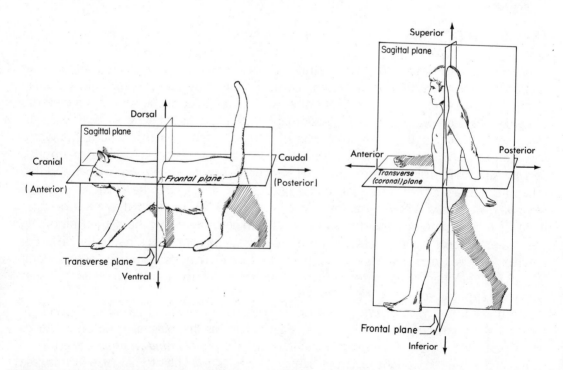

Figure N–1 Diagrams to show the planes of the body and the differences in the terms for direction in a quadruped and in a human being.

directions are referred to as *posterior* and *anterior.* It is recommended that the terms anterior and posterior not be used in a quadruped for the cranial and caudal ends of the body because of the possible confusion with the different usage in human anatomy, but anterior and posterior may be used for structures in the head and appendages, for these relationships are essentially the same in quadrupeds and human beings.

Other terms for direction are used in the same way in all animals. *Lateral* refers to the side of the body; *medial* to a position toward the midline. *Median* is used for a structure in the midline. *Distal* refers to a part of some organ, such as an appendage or blood vessel, that is farthest removed from the point of reference, such as the center of the body or the origin of the vessel; *proximal,* to the opposite end of the organ, i.e., the part nearest the point of reference.

Left and *right* are self-evident, but it should be emphasized that in anatomical directions they always pertain to the *specimen's* left or right, regardless of the way the specimen is viewed by the observer.

Adverbs may be formed from the above adjectives by adding the suffix *-ly* or *-ad* to the root, in which case the term implies motion in a given direction. To say that a structure extends caudally, or caudad, means that it is moving toward the tail.

Planes and Sections of the Body

A body is frequently cut in various planes to obtain views of internal organs. A longitudinal, vertical section from dorsal to ventral that passes through the median longitudinal axis of the body is a *sagittal* section. Such a section lies in the sagittal plane. Sections or planes parallel with, but lateral to, the sagittal plane are said to be *parasagittal.*

A section cut across the body from dorsal to ventral, and at right angles to the longitudinal axis, is a *transverse* section, and it lies in the transverse plane.

A *frontal* (coronal) section or plane is one lying in the longitudinal axis, and passing horizontally from side to side.

Chapter One

THE EVOLUTION AND EXTERNAL ANATOMY OF MAMMALS

Although this manual is designed for the study of the cat, certain comparisons will be made with other mammals and with lower vertebrates. This is done to place mammalian anatomy in an evolutionary perspective and thereby make it more meaningful. There is much about the structure of mammals, human beings included, that is best understood by referring to earlier stages of vertebrate evolution. It is desirable, therefore, to begin with a brief review of vertebrate evolution.

VERTEBRATE EVOLUTION

The subphylum *Vertebrata* is divided into eight classes, four of which are aquatic (fishes) and four essentially terrestrial (tetrapods). Their relationships are depicted in Figure 1–1. As can be seen, the ancestral class was the *Agnatha.* This is an ancient group, the earliest fossils being found in geologic deposits nearly 500,000,000 years old. The class consists of several extinct orders of heavily armored fishes, collectively known as ostracoderms, and the living lampreys and hagfishes. All are characterized by the absence of jaws, and most lack paired appendages.

Vertebrates above the Agnatha have jaws and typically paired appendages. Having jaws, they are often referred to as *gnasthostomes.* Among the earliest gnathostomes were members of the class *Placodermi.* Their jaws and paired appendages were rather primitive and also variable among the widely divergent fishes that compose the class. Unhappily, none has survived.

Living sharks and skates of the class *Chondrichthyes* probably evolved from early placoderms. They have well developed jaws and two pairs of paired appendages, and they are distinguished from other living fishes by skeletons that are entirely cartilaginous and by a lack of lungs or swim bladders.

All remaining fishes belong to the class *Osteichthyes.* They evolved either from early placoderms or possibly directly from ostracoderms. They too have well-developed jaws and paired appendages but differ from the cartilaginous fishes by having at least partially ossified skeletons and generally lungs, or lung derivatives (swim bladder). The class is a very large one and includes the vast majority of living fishes—sturgeon,

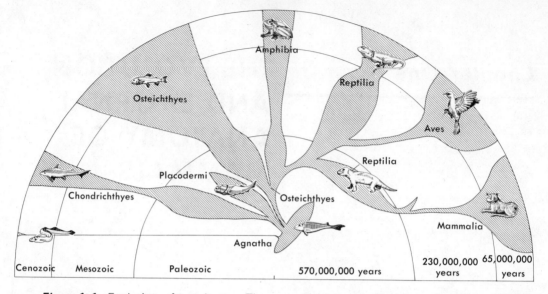

Figure 1–1 Evolution of vertebrates. The presumed interrelationships of the eight classes of vertebrates, their distribution in time, and their relative abundance are shown.

garpike, eels, minnows, perch, bass, etc. The lungfishes and their close relatives, the crossopterygians, are also an evolutionary branch of this class.

Freshwater crossopterygians gave rise to tetrapods, the earliest of which were cumbersome amphibians (class *Amphibia*) collectively known as *labyrinthodonts.* Some fossils are 350,000,000 years old. Labyrinthodonts, in turn, gave rise directly or indirectly to the reptiles and to the living amphibians: frogs and toads, salamanders, and wormlike caecilians of the tropics. Like living amphibians, the extinct labyrinthodonts were presumably transitional between an aquatic and a terrestrial mode of life. They had acquired legs and other terrestrial adaptations but were found only close to water. Most amphibians live near fresh water, for their ability to conserve body water is rudimentary, and most reproduce in this medium since they have not evolved a cleidoic egg with its large store of yolk, extraembryonic membranes, fluid, and shell that provide for all the requirements of the developing embryo, and thereby obviate an aquatic larval stage.

Reptiles (class *Reptilia*) have completed the transition from water to land by evolving such an egg and the ability to conserve body water. Reptiles and higher classes are often called *amniotes,* the amnion being a fluid-filled extraembryonic membrane that surrounds the embryo; amphibians and fishes are *anamniotes.* Although living reptiles resemble anamniotes in being dependent upon external sources for most of their body heat (*ectothermic*), they can maintain a relatively high body temperature when they are active by behavioral regulation of their degree of exposure to the sun. Certain extinct groups may have had some degree of physiological control of body temperature. Reptiles were the dominant terrestrial vertebrates for 200,000,000 years and during this period adapted to many habitats. The dinosaurs became giants of the land and swamps. Other reptiles evolved true flight, and some returned to the sea. Of the numerous reptilian orders only the turtles, *Sphendon,* lizards and snakes, and crocodiles and alligators have survived. Although these groups are reasonably successful, mammals and birds now occupy a position of dominancy.

Both birds and mammals improved on the basic terrestrial adaptations of reptiles, gaining the ability to maintain a high and relatively constant body temperature by physiological regulation (*endothermic*). This enabled them to be active under a wide range of external temperatures. Birds (class *Aves*), as a group, specialized for flight, evolving wings and feathers. Primitive birds had toothed jaws, clawed fingers, and long tails and, aside from their feathers, were very similar to, and probably evolved from, certain small bipedal reptiles closely related to early dinosaurs. More recent birds have lost these reptilian features but continue to lay reptilian type eggs.

The line of evolution to mammals (class *Mammalia*), which diverged early from that to other reptiles, passed through two extinct reptilian orders collectively referred to as the *mammal-like reptiles.* As the name implies, these reptiles gradually came to resemble mammals; indeed, some of the later species probably were endothermic and covered by an insulating layer of hair. The transition from mammal-like reptile to mammal included certain changes in the nature of the jaw joint to make it more efficient for mastication and the evolution of a middle ear with three auditory ossicles rather than one. The first mammals appeared about 200,000,000 years ago, even before the complete dominancy of the reptiles, but mammals remained inconspicuous in the fauna for millions of years. The most primitive of living mammals are the duckbilled platypus *(Ornithorhynchus)* and spiny anteater *(Tachyglossus)* of the Australian region, belonging to the subclass *Prototheria,* order *Monotremata.* Like other mammals, they have hair, mammary glands, although no nipples, and the characteristic type of jaw joint and auditory ossicle mechanism. But they continue to lay a reptilian type of egg. Although practically unknown as fossils, monotremes must be an ancient group.

Other living mammals are placed in the subclass *Theria* and reproduce viviparously; that is, they retain the embryo in a uterus where its needs are provided for by some sort of placenta until it is ready for birth. With the extinction of the dominant reptilian orders about 65,000,000 years ago, mammals came into their own. Two lines of evolution diverged from primitive therians. One led to the infraclass *Metatheria,* which includes but one order, the *Marsupialia.* Marsupials are pouched mammals such as the kangaroo of Australia and the opossums of America. They generally have a poorly formed placenta and so give birth to young at a relatively early stage. The young move into the pouch, or marsupium, where they attach to nipples of the mammary glands and complete their development. The other line of evolution led to the infraclass *Eutheria,* or true placental mammals. Eutherians lack the marsupium, and have a more efficient placenta, for the embryos are retained in the uterus until a much more advanced stage. They are now the dominant mammals. The order *Insectivora,* which includes the shrews and moles, is the most primitive of the numerous orders, and the others diverged directly, or indirectly from it. Human beings, *Homo sapiens,* belong in the order *Primates;* cats and other flesh-eating mammals, to the order *Carnivora.* The generic and specific name of the domestic cat is *Felis catus.*

EXTERNAL ANATOMY

Aside from the skin, most of the external features of vertebrates are simply manifestations of other organ systems. Nevertheless, it is worthwhile to study them along with the skin at this time, for this will bring out general changes that have occurred during evolution in body shape and regions, appendages, openings, and special sense organs.

General External Features

Examine a cat (Fig. 1–2). The diagnostic *hair* of mammals is at once evident. It will also be recognized that body regions are more clearly defined than they are in fish and the more primitive terrestrial vertebrates. This correlates primarily with a change in methods of locomotion—from swimming to effective walking and running. The *head (corpus)* is large and separated from the *trunk* by a distinct and movable *neck (collum)*. The trunk itself can be divided into *back (dorsum)*, *thorax*, encased by the ribs, *abdomen,* and *pelvis*. A *tail (caudal)* is typically present in mammals but, except in the whales and their allies, is greatly reduced in size in comparison with that of a primitive tetrapod. In some terrestrial mammals it is of use, often as a balancing organ in semiarboreal canivores; in some other mammals it is a small organ used as a warning signal (rabbit); and in a few (human beings) it has been lost as an external structure.

The paired appendages consist of a *brachium,* an *antebrachium,* and a *manus,* in the pectoral appendage; a *femur*[1], a *crus* and a *pes,* in the pelvic appendage. Observe either on a mounted specimen, or on a skeleton, that a cat walks on its toes with the wrist and heel raised off the ground. This method of locomotion is referred to as *digitigrade,* in contrast to *plantigrade* (human beings), in which the entire sole of the foot is flat on the ground, or *unguligrade* (ungulates such as the horse), in which the animal walks on the tip of its toes. The most medial, or first, toe would not reach the ground in a digitigrade mammal, and it is vestigial in the manus of a cat and completely lost in the pes as an externally visible structure. The terminal segment of each toe has a *claw*, and this segment is hinged in such a way that the claw can be retracted or extended.

1. The term femur can be used both for the thigh and for the bone within it. The **Nomina Anatomica** term for the thigh is femur, and for the bone, it is os femoris.

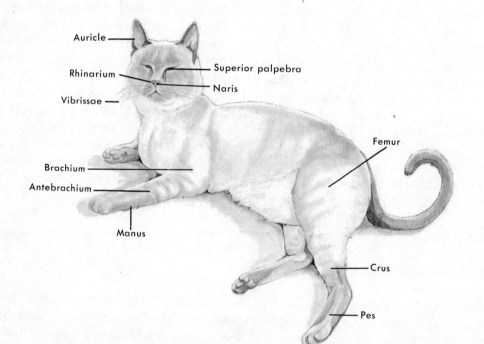

Figure 1–2 External features of a cat, *Felis catus.*

In primitive terrestrial vertebrates, such as a frog or salamander, the brachium and femur move back and forth close to the horizontal plane. This is a primitive and rather awkward method of support and movement. During the evolution of mammals, the limbs have rotated in such a way that they are beneath the body and the legs move back and forth nearly in the vertical plane. In the hind leg, the change results from a 90-degree forward rotation. The knee and pes point anteriorly, and the original preaxial, or anterior, border is medial instead of anterior. In the front leg, there has been a 90-degree backward rotation, so that the elbow points posteriorly. Because of a continued torsion at the elbow, seen beginning in primitive tetrapods, the manus still points anteriorly. The original preaxial border of the brachium is now lateral instead of anterior, but that of the antebrachium continues to be primarily medial. In order to understand these changes, try to visualize your own appendages in the primitive tetrapod position, and then slowly rotate them into the quadruped mammal position. The new limb posture is more efficient for support. It also increases speed, for the legs are now capable of a long and rapid fore and aft swing. Walking on the toes, with the wrist and heel raised, increases the length of the stride. Although the trunk and tail are flexible, lateral undulations play no significant part in the locomotion of mammals, as they do in fishes and primitive terrestrial vertebrates. Other changes have occurred in the limbs of human beings, as our ancestors adapted to a bipedal gait (cf. chapter 3).

Turning the head, notice that the *mouth (os)* is bounded by fleshy *lips (labia)*. The paired *external nostrils (nares)* are close together on the *nose* and are surrounded by moist, bare skin known as the *rhinarium.* The *eyes (oculi)* are large and protected by movable upper and lower *eyelids (palpebrae).* Spread these lids apart, and observe a third lid, called the *nictitating membrane,* in the medial corner of the eye. The nictitating membrane can be drawn across most of the eye, thus helping to moisten and cleanse this organ. It is reduced to a vestigial *semilunar fold* located in the medial corner of the human eye. Mammals have a prominent external ear consisting of a conspicuous external flap, called the *auricle,* and an external ear canal *(external acoustic meatus)* that extends into the head from the base of the auricle. The eardrum *(tympanum)* is located at the bottom of the meatus. It will not be seen at this time. That part of the head that includes the jaws, mouth, nose, and eyes is referred to as the *facial region;* the rest, containing the brain and ears, is the *cranial region.*

Primitive vertebrates have a common chamber, the cloaca, which receives the products of the digestive, urinary and reproductive tracts. In therian mammals, the cloaca has become divided so that the intestine and urogenital ducts open independently at the surface. The opening of the intestine, called the *anus,* will be found just ventral to the base of the tail. In most female mammals, the combined opening of the urinary and reproductive ducts will appear as a second opening (the *vaginal vestibule*) bounded by small folds, ventral to the anus. The vaginal vestibule is a relatively long canal in quadrupeds, but the comparable region of a woman, known as the *vulva,* is shallow, for the vagina and the urethra are independent structures almost to the body surface. In male mammals, the urogenital ducts open at the tip of a *penis.* A sac-shaped *scrotum* containing the testes lies caudal to the penis. The entire area of anus and external genitals is called the *perineum* in both sexes. Further discussion of the details of the external genital organs will be deferred.

Carefully feel along the ventral surface of the thorax and abdomen on each side of the midline, and you will find two rows of teats *(papillae mammae)* hidden in the fur. These bear the minute openings of the mammary glands. The teats are more prominent in females, but rudiments can sometimes be found in males. There are about five pairs in the cat, but the number is subject to variation. We have a single pectoral pair.

INTEGUMENTARY DERIVATIVES

The *integument (cutis)* of all vertebrates consists of two fundamental layers of tissue—an outer *epidermis* of stratified epithelial cells derived from the embryonic ectoderm, and an inner *dermis (corium)* of dense connective tissue derived from the mesoderm (Fig. 1–3). The dermis is the thicker and also the more stable layer, for its basic structure has remained much the same during vertebrate evolution. But as vertebrates adapt to a terrestrial environment, the epidermis becomes thicker, and its outer cells become cornified (keratinized); that is, the outer cells, as they die, become filled with a horny, water insoluble protein called *keratin* that renders them less pervious to water. The cornified cells, which often form a distinct epidermal layer called the *stratus corneum,* are continually being sloughed off, but they are replaced by mitosis in the basal cells of the epidermis, the *stratum basale.* Transitional layers between the stratum basale and the corneum are often recognized.

These changes cannot be seen grossly in the laboratory, but certain derivatives of the integument can be studied along with the general external features. The integument of mammals is rich in derivatives. Chromatophores, which contain the pigment in lower vertebrates, are largely absent, but pigment granules are present in and between the epithelial cells. The numerous glands, which are derived from epidermal cells, fall into three categories—*mammary glands,* tubular-shaped *sweat glands (sudoriferous glands),* and avleolar-shaped *sebaceous glands* usually associated with the hair follocles. The mammary glands resemble sweat glands in having contractile myo-epithelial cells peripheral to the secretory cells and, for this reason, are considered to be modified sweat glands by some investigators. There are two types of sweat glands, apocrine and eccrine. *Apocrine glands,* which are abundant in the armpits and in the genital areas, discharge into the hair follicles and produce secretions responsible for body odors.

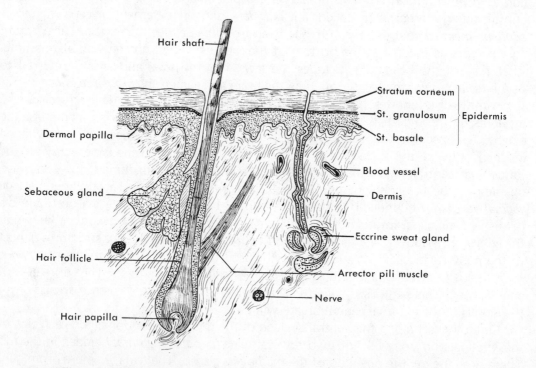

Figure 1–3 Diagram of a vertical section through the skin of a mammal.

Eccrine glands secrete a more watery solution that is of value in cooling the body. The functions of the sebum, which is discharged by the sebaceous glands into the hair follicles, is not entirely clear; possibly it helps to maintain the condition of the hair. Some sebaceous glands are modified into wax glands. The openings of the eccrine sweat glands can be seen with a hand lens on one's finger tips; they are more abundant in human beings than in heavily furred mammals. The openings of the sebaceous glands and hair follicles are the more familiar "pores" of the skin.

The most conspicuous integumentary derivative is the insulating covering of *hair (capillus)*. Hair replaces the horny scales of reptiles in most mammals, but scales may still be found on the tails of certain rodents, and they have redeveloped over the bony plates of the armadillo shell. Although hair is composed of keratinized cells, it is a new development of the epidermis. It is not considered to be homologous with either horny scales or feathers, for details of its embryonic development are different. Moreover, the distribution of hair, as seen, for example, on the back of one's hand, leads to the conclusion that hairs evolved in small clusters between the scales, and then the scales were lost. The simultaneous presence of scales and hairs can be seen on the tails of some rodents. In most mammals, the hair forms a dense fur over the body, being modified in certain places such as the *eyelashes (cilia)* and *tactile whiskers (vibrissae)* on the head of the cat. But there are many departures from this pattern. Hair is reduced in human beings, and lost in the adults of such highly aquatic mammals as the whale. In some other mammals, the hair has become adapted for very specialized purposes. The quills of a porcupine are a case in point, and the "horn" of a rhinoceros resembles a compact mass of hair.

Reptilian *claws* are retained in most mammals but have been transformed into *nails (ungues)* in certain primates, and into *hoofs* in the ungulates. Other common integumentary derivatives are the *foot pads (tori)* on the feet of most mammals. These are simple thickenings of the stratum corneum. The whorls of *friction ridges* (fingerprints) of primates are their homologue.

Aside from the widely distributed structures mentioned above, some mammals have still other derivatives. The redevelopment of *bone* in the dermis of the armadillo, the keratinized *whalebone plates* within the mouth of the toothless whales, and the horny covering of the *horns* of sheep, antelope and cattle belong in this category. The horns of these animals consist of a core of bone which arises from the skull and is covered with very heavily keratinized epidermis, i.e., horn. Horns of this type should not be confused with *antlers* found in the deer group. Antlers are bony outgrowths of the skull that are covered with skin (the velvet) only during their growth. In contrast with horns, they generally are restricted to the male; they branch; and they are shed annually.

Chapter Two
THE AXIAL AND VISCERAL SKELETON

The next organ systems to be studied will be a group involved with the general functions of support and locomotion, namely, the skeleton, muscles, sense organs, and nervous system. The skeleton may appropriately be considered first, as it is a fundamental system about which the body is built.

DIVISIONS OF THE SKELETON

The vertebrate skeleton consists of two basic parts—the *dermal skeleton* and the *endoskeleton*. Although these two become united in various degrees, they are distinct in their ontogenetic and phylogenetic origins. The dermal skeleton consists of bone that develops embryologically directly from the mesenchyme in, or just beneath, the dermis of the skin. This type of bone is called either *dermal* or *membrane bone*. It follows that the dermal skeleton is superficial. The bony scales and plates of fishes, and their derivatives, are dermal in nature. Among their derivatives in mammals are the dermal plates in the skull, teeth, and the dermal portions of the pectoral girdle. In addition, dermal bone has evolved independently of bony scales in the dermis of certain animals. Portions of the shell of the turtle and the armadillo are familiar examples.

The endoskeleton, on the other hand, arises in deeper body layers, and consists of cartilage or bone that develops in association with cartilaginous rudiments. Although such bone has the same histological structure as dermal bone, it is convenient to differentiate it as *cartilage replacement bone*. The endoskeleton may be subdivided into visceral and somatic portions. The *visceral skeleton,* as the name implies, is associated with the "inner tube" (gut) of the body. It consists of skeletal arches (visceral arches) that form in the wall of the pharynx, and support the gills in fishes. The *somatic* portion of the endoskeleton is associated with the "outer tube" of the body (body wall and appendages). It may be further broken down into axial and appendicular subdivisions. The *axial skeleton* includes those parts of the somatic skeleton located in the longitudinal axis of the body—vertebrae, ribs, sternum, and those portions of the brain case composed of cartilage or cartilage replacement bone (chondrocranium). The *appendicular skeleton* consists of the more laterally placed portions of the somatic skeleton—the skeleton of the paired appendages and those portions of their girdles composed of cartilage or cartilage replacement bone. The divisions of the different parts of the skeleton summarized in Table 2–1.

One must be continually aware of the difference between the endoskeleton and the dermal skeleton, but in studying the evolution of the entire skeleton, this dichotomy cannot always be followed. It is often easier to consider the dermal skeleton along with those portions of the endoskeleton with which it becomes associated. In approaching the endoskeleton, it is convenient to

TABLE 2-1　*Divisions of the Skeleton*

Dermal Skeleton (Dermal Bone)	Endoskeleton (Cartilage Replacement Bone)
Bony Scales Dermal Plates (become associated with parts of endoskeleton) Teeth	Somatic Skeleton (in body wall) 　Axial Skeleton (chondrocranium, vertebrae, ribs, sternum) 　Appendicular Skeleton (girdles, bones of paired appendages) Visceral Skeleton (in gut wall) 　Visceral Arches

study the axial and visceral portions together to some extent because they are combined in the formation of the skull. After their evolution has been traced, that of the appendicular skeleton will be followed.

The following directions are based on the cat, but they could be applied to many other mammals. If material is available, it would be desirable to compare the skeleton of the cat with the human skeleton and that of other species.

POSTCRANIAL AXIAL SKELETON

The mammalian postcranial axial skeleton is far more complex than that of lower vertebrates. In a fish, for example, much of the weight of the body is supported by the surrounding water, so the vertebral column and ribs are essentially a girder on which the segmented trunk and tail muscles act in bringing about the undulatory locomotor movements. The individual vertebrae are not much different from one another—only trunk and tail vertebrae can be easily distinguished. Nearly all bear ribs. In terrestrial vertebrates, body weight must be supported by the vertebral girder, head and tail move independently of the trunk, trunk movements are more complex, and in most reptiles and all mammals the ribs play an important role in respiratory movements. All this necessitates an exceptionally strong vertebral column with interlocking vertebrae and considerable differentiation of the vertebrae and ribs in different parts of the body.

(A) VERTEBRAL GROUPS

Examine a mounted skeleton of the cat or human being, and a set of disarticulated vertebrae. There are five groups of vertebrae in mammals. The most cranial group is the neck or *cervical vertebrae.* There are more of them than in very primitive tetrapods, for, with few exceptions, mammals have seven cervical vertebrae, all of which lack ribs moveably articulated to the vertebrae. *Thoracic vertebrae* bearing ribs follow the cervical, and *lumbar vertebrae* follow the thoracic. The lumbar vertebrae lack ribs but have very large transverse processes. The number of vertebrae in these two regions varies among mammals. The cat normally has 13 thoracic and seven lumbar vertebrae; we have 12 thoracic and five lumbar vertebrae. *Sacral vertebrae* follow the lumbar and are firmly fused together to form a solid point of attachment, the *sacrum,* for the pelvic girdle. The number of vertebrae contributing to the sacrum varies between species. Most mammals, including the cat, have three; we have five, for the sacrum must be very strongly attached to the pelvic girdle in a biped. The remaining are *caudal vertebrae* (coccygeal vertebrae). Their number varies with the length of the tail, but all mammals have some. Even in human beings there are three to five small coccygeal vertebrae that are fused together forming a *coccyx* to which certain anal muscles attach.

(B) THORACIC VERTEBRAE

The thoracic vertebrae should be studied first since they are less specialized than those of other regions. Examine one from near the middle of the series (Fig. 2–1). Identify the *vertebral arch* with its long *spinous process.* The spinal cord is encased by the vertebral arches and lies within the *vertebral canal.* The large, solid *vertebral body* or *centrum* lies ventral to the spinal cord and forms the chief supporting element of a vertebra. Each end of the body is flat, a shape termed *acelous.* In life, small, fibrocartilaginous *intervertebral discs* are located between successive bodies. Articulate two vertebràe and notice how the back of one vertebral arch overlaps the front of the one behind it. Disarticulate the vertebrae and look for smooth articular facets in the area where the vertebral arches come together. These are the articular processes (zygapophyses). The facets of the pair of *cranial articular processes* are located on the cranial surface of the vertebral arch and face dorsally; those of the pair of *caudal articular processes* are on the caudal surface of a vertebral arch and face ventrally. A leteral transverse process, known as a *diapophysis,* projects from each side of the vertebral arch. Notice the smooth *articular facet* for the tuberculum of a rib on its end. In the cranial and middle portion of the thoracic series, the head of a rib articulates between vertebrae, so part of the facet for a head is located on the front of one vertebral body, and part on the back of the next cranial body. The somewhat constricted portion of the vertebral arch between the diapophysis and body is called its *pedicle.* When several vertebrae are articulated, you will see holes for the spinal nerves, the *intervertebral foramina,* between successive pedicles.

Compare one of the thoracic vertebrae from the middle of the series with those near the cranial and caudal ends of the thoracic region and observe that the spinous processes, articular processes, transverse processes, and the position of the facets for rib heads differ in details. For example, the most caudal thoracic vertebrae lack transverse processes, for their ribs lack a tuberculum, and the facets for their rib heads are entirely on one vertebral body. Articulate the last two thoracic vertebrae and notice how the articulation of the articular processes is reinforced in the cat by a process of the pedicle that extends caudad lateral to the articular processes. This is an *accessory process;* it is absent in human beings. A skilled observer can find differences among the vertebrae sufficient to identify each one precisely as the third, sixth, or eighth thoracic, etc. However, it will be sufficient for students to be able to distinguish the major group to which a vertebra belongs. All thoracic vertebrae have at least one articular facet for a rib. No other vertebrae have such facets.

(C) LUMBAR VERTEBRAE

Lumbar vertebrae are characterized by their large size and by having prominent, bladelike transverse processes. The transverse process is composed of a vertebral process (diapophysis) to which an embyonic rib has fused; hence, it is called a *pleurapophysis.* The lumbar vertebrae of the cat also have a small bump for the attachment of ligaments and tendons situated dorsal to the articular surface of each

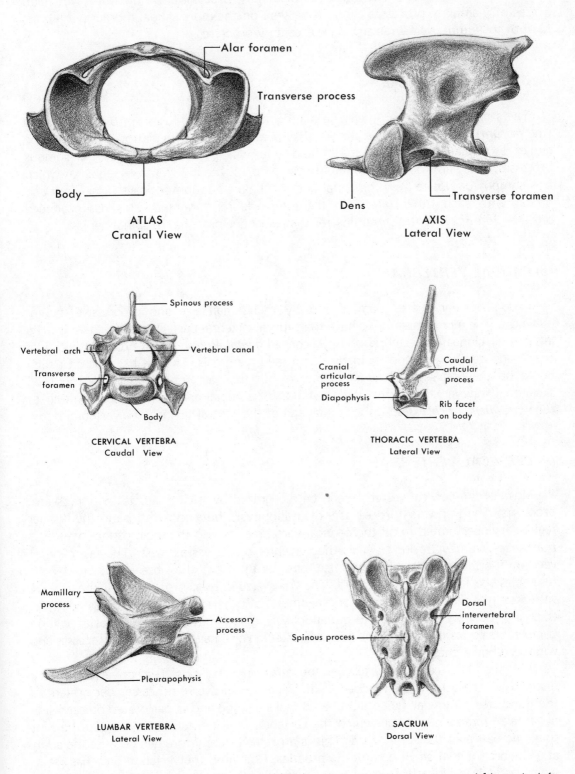

Figure 2–1 Drawings of selected vertebrae of the cat. In the lateral views, *cranial* is to the left of the drawing.

cranial articular process. These are the *mamillary processes;* we do not have them. Traces of mamillary processes can also be seen on the more caudal thoracic vertebrae. Most of the lumbar vertebrae also have accessory processes.

(D) SACRAL VERTEBRAE

The sacral vertebrae can easily be distinguished by their fusion into a single piece (the *sacrum*) and by the broad surface they present for the articulation with the pelvic girdle. Examine the sacrum closely and you will be able to detect the spinous processes, articular processes, pleurapophyses, etc., of the individual vertebrae of which it is composed. Notice how the distal ends of the pleurapophyses have fanned out and united with each other lateral to the intervertebral foramina. This has produced separate dorsal and ventral foramina for the respective branches of spinal nerves.

(E) CAUDAL VERTEBRAE

Caudal vertebrae are characterized by their small size and progressive incompleteness. The more cranial ones have the typical vertebral parts, but soon there is little left but an elongated vertebral body. V-shaped *hemal arches,* which protect the caudal artery and vein, are found in many quadruped mammals. The cat has traces of such bones on the first several caudal vertebrae, but they are usually lost in a mounted skeleton. However, the points of articulation of a hemal arch will appear as a pair of tubercles, *hemal processes,* at the cranial end of the ventral surface of a vertebral body.

(F) CERVICAL VERTEBRAE

Most of the cervical vertebrae can be recognized by their characteristic transverse processes. The transverse process is a pleurapophysis, but, unlike those in the lumbar region, it is perforated in all the cervical vertebrae, except the most caudal one, by a *transverse foramen* through which the vertebral blood vessels pass. The last cervical vertebra normally lacks this foramen and, aside from the absence of rib facets, resembles the first thoracic vertebra. The transverse processes of many of the cervical vertebrae (the fifth or sixth is a good example) also have two parts—a dorsal, pointed transverse portion comparable to a diapophysis, and a ventral, platelike costal portion comparable to a rib. Most of the cervical vertebrae also have low spinous processes and wide vertebral arches.

The first two cervical vertebrae, the *atlas* and the *axis,* respectively, are very distinctive. The atlas is ring-shaped, with winglike transverse processes perforated by the transverse foramina. Its vertebral arch lacks a spine and is perforated on each side by an *alar foramen,* through which the vertebral artery enters the skull and the first spinal nerve leaves the spinal cord. This is an intervertebral foramen because the adult atlas incorporates a small embryonic proatlas. Cranially, the vertebral arch has facets which articulate with the two occipital condyles of the skull; caudally, it has facets that articulate with the body of the axis. The body of the atlas is reduced to a thin transverse rod.

The axis is characterized by an elongated spinous process that extends over the vertebral arch of the atlas; very small transverse processes; rounded articular surfaces at the cranial end of its body; and a median, tooth-like *dens* that projects from the front of the body of the axis into the atlas.

Although an atlas is present in amphibians, an axis does not appear until reptiles. Together these two vertebrae form a universal joint which permits the free movement of the head characteristic of amniotes. The composition of the atlas and axis is also somewhat different from that of other vertebrae, for part of the centrum of the atlas has united with that of the axis and forms the dens. This explains the relatively small size of the body of the atlas and the projection of the dens into the atlas.

(G) RIBS

Study the *ribs (costae)* from disarticulated specimens and on a mounted skeleton. There are 13 in the cat and 12 in human beings. Most of them have both heads characteristic of tetrapods—a proximal *head* articulating with the vertebral body and a more distal *tuberculum* articulating with the transverse process (Fig. 2–2), but the last three (cat) or two (human) ribs have only the head. The portion of the rib between its two heads is its *neck;* the long distal part, its *body* (shaft). A *costal cartilage* extends from the end of the body. Those ribs whose costal cartilages attach directly on the sternum are called *vertebrosternal ribs;* those whose costal cartilages unite with other costal cartilages before reaching the sternum are *vertebrocostal ribs;* those whose costal

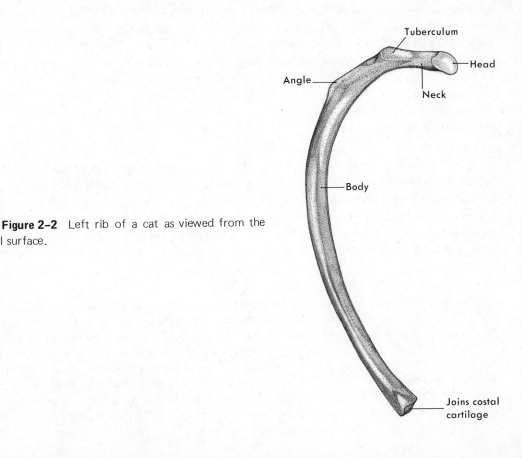

Figure 2–2 Left rib of a cat as viewed from the caudal surface.

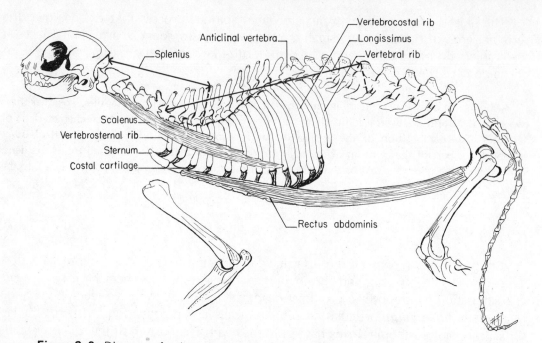

Figure 2-3 Diagram of a lateral view of a cat's skeleton to illustrate Slijper's views on the biomechanics of the trunk. The scapula has been omitted. Heavy arrows indicate the line of action of certain muscles upon the spinous processes. (Modified after Slijper.)

cartilages have no distal attachment are **vertebral ribs** (Fig. 2–3). Sometimes vertebrosternal ribs are referred to as **true ribs** and all the rest are referred to as **false ribs.** A floating rib is an alternate name for a vertebral rib. The cat normally has nine vertebrosternal, three vertebrocostal, and one vertebral rib; we have five vertebrosternal, five vertebrocostal, and two vertebral ribs.

(H) STERNUM

The **sternum** of the cat is composed of a number of separate, ossified segments called the **sternebrae.** The first of these constitutes the **manubrium;** the last, the **xiphisternum;** and those between are the **body** of the sternum. A cartilaginous **xiphoid process** extends caudad from the xiphisternum. Costal cartilages join between successive sternebrae. We have a distinct manubrium to which the clavicle attaches, but the remaining sternebrae fuse during embryonic development to form a unified sternal body.

The vertebral column of mammals is the primary girder of the skeleton. Not only does it carry the body weight and transfer this to the appendages, but, through its extension and flexion, it participates in the locomotor movements of many quadrupeds. In order to facilitate interpreting its structure, some authors have compared the vertebral column of a quadruped to a cantilever girder, but Slijper (1946) and others have taken a broader view and have compared the entire trunk skeleton and associated muscles to a bowstring bridge, or to an archer's bow (Fig. 2–3). The vertebral bodies represent the main supporting arch, but this is a dynamic arch that has a tendency to straighten out because of the elasticity of certain dorsal ligaments and the tonus of the dorsal muscles. This is prevented by the "bow string"—the sternum and ventral abdominal muscles such as the rectus abdominis. The "bow" and "string" are connected caudally by the pelvic girdle and cranially by the stout cranial ribs, which are held in place by the action of such muscles as the scalenus. This hypothesis has the merit of viewing the vertebral column as a dynamic girder capable both of providing support and of participating in the movements of the body.

The spinous processes are viewed as muscle lever arms acting on the vertebral bodies. The direction of their inclination tends to be perpendicular to the major muscle forces acting upon them. Since several muscles may attach onto a single spine, the situation becomes quite complex, but in carnivores and many other mammal quadrupeds the tips of the spines of the lumbar vertebrae point toward the head, partly in response to a very powerful longissimus muscle (Fig. 4–24, p. 75), whereas the spines of the thoracic vertebrae slope in the opposite direction, partly in response to a powerful splenius. The vertebra near the middle of the trunk, where the angle of inclination of the spines reverses, is called the *anticlinal vertebra.*

If the spinous processes are lever arms, an increase in their height increases the length of the lever arms and the mechanical advantage of muscles acting upon them. The great length of the anterior thoracic spinous processes is considered to be related to the support of a relatively heavy head by the splenius and other muscles.

In man, a biped, the vertebral column is a vertical beam with the head balanced upon the top. Forces acting on the column are quite different than in a quadruped. The spinous processes of human vertebrae are not so long as a cat's, there is no anticliny, and the vertebral bodies become progressively larger toward the sacrum where all the weight of the upper part of the body is transferred to the pelvic girdle and hind legs. We also have a lumbar curve which shifts the center of gravity over the pelvic girdle and hind legs.

HEAD SKELETON

The head skeleton of mammals, which comprises the skull proper, lower jaw and hyoid apparatus, has had a complex evolutionary history. In fishes ancestral to tetrapods, and in primitive terrestrial vertebrates, one can recognize three rather distinct components at the head skeleton (Fig. 2–4). (1) A *chondrocranium,* which is the cranial portion of the axial skeleton, more or less surrounds the brain (except much of the top of the brain), and encapsulates the inner ear and olfactory sac. Like the vertebrae, it is composed of cartilage or cartilage replacement bone. (2) A series of seven *visceral arches* lie for the most part ventral and caudal to the chondrocranium. The first of these, the mandibular arch, forms the "core" of the upper and lower jaws and bears the jaw joint. The portion of the mandibular arch associated with the upper jaw is called the palatoquadrate. The dorsal portion of the second visceral arch (hyoid arch) is known as the *hyomandibular cartilage,* and it extends from the otic capsule of the chondrocranium to the caudal end of the palatoquadrate cartilage, which it helps to support. In ancestral tetrapods, the hyomandibular cartilage probably also made contact with an eardrum. All visceral arch elements are also cartilage or cartilage replacement bone. (3) Finally, the chondrocranium and visceral arches are encased by *dermal bone.* A dermal roof covers the brain dorsally and extends laterally and ventrally across the jaw muscles and over the palatoquadrate. Teeth occur on much of its ventral margin, and it is perforated by openings for the external nostrils and eyes. Palatal dermal bones occur in the roof of the mouth on the ventral side of the palatoquadrate and chondrocranium. They constitute the primary palate, which in tetrapods is perforated anteriorly by a pair of internal nostrils. Dermal bones encase the ventral portion of the mandibular arch. Teeth in both the upper and lower jaws are carried on dermal bones but the jaw joint itself is borne by the mandibular arch. The remaining visceral arches are covered laterally and ventrally by dermal opercular and gular bones.

The touchstone of the evolution of mammals from primitive terrestrial vertebrates has been the development of an increased level of activity. Mammals are very active, endothermic vertebrates that can maintain a relatively high and constant body temperature and level of metabolism. This would

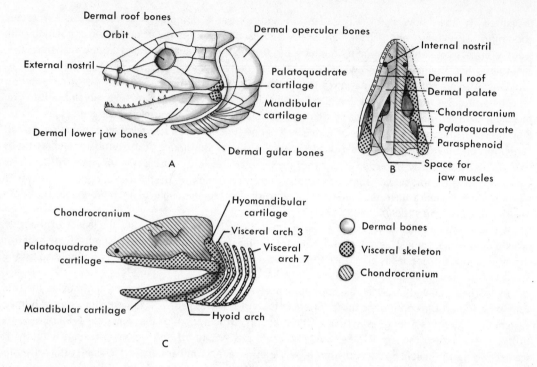

Figure 2–4 Components of the vertebrate head skeleton as seen in a generalized fish. *A*, Lateral view, showing superficial dermal bones covering most of the deeper elements; *B*, ventral view of the skull, with dermal bones removed on the right side of the drawing; *C*, lateral view after removal of the dermal bones.

not be possible without concomitant structural changes in all organ systems. The skull has been affected primarily by the great enlargment of the brain needed in an active animal and by changes in feeding mechanisms that make possible the increased intake and processing of food.

One important change was a great expansion of the braincase (Fig. 2–5). In the mammalian braincase, chondrocranial elements are more or less confined to the back and underside of the

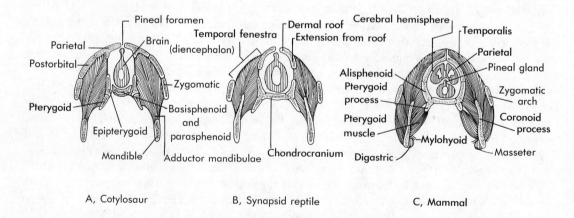

Figure 2–5 Evolution of the vertebrate skull as seen in cross sections caudal to the orbit of an ancestral reptile (*A*), a mammal-like reptile (*B*), and a mammal (*C*). Development of the temporal fenestrae, the jaw musculature, and the braincase is shown.

braincase. The brain is largely encased by processes of dermal bone that extended inward from the original dermal roof. It can be seen from the figure that the original roof lay lateral to the jaw muscles and that these processes are medial to the muscles. Although derived from the roof, these dermal processes do not constitute the original roof. The braincase is completed laterally by the epipterygoid, which becomes the wing of the basisphenoid (alisphenoid). The epipterygoid is derived from that part of the palatoquadrate cartilage that in more primitive vertebrates forms a movable articulation between the palatoquadrate cartilage and the chondrocranium.

As the braincase expanded and the jaw muscle enlarged during the evolution of mammals, the space for the temporal jaw muscles, which lies between the temporal portion of the dermal roof and the braincase, became reduced (Fig. 2–5). This situation was met by the loss of some of the dermal roof on each side, thereby forming a window, or *temporal fenestra,* through which the muscles could bulge, especially during their contraction. Temporal fenestrae make their appearance in the mammal-like reptiles and are of the *synapsid* type; that is, there is only one on each side located on the ventrolateral portion of the skull. At first the fenestra was small, but it gradually enlarged. In primitive mammals most of the original lateral surface of the dermal roof has disappeared in the temporal region, so the temporal muscles are lodged in a large *temporal fossa* that merges with the orbit. What appears to be the lateral dermal roof in a mammal skull is the extension of the dermal bones contributing to the braincase. All that is left of the original dermal roof in this region is a strip of bone in the middorsal line, another bordering the occipital region, and the handlelike *zygomatic arch* lying ventral to the temporal fossa and orbit. Temporal muscles pass medial to the zygomatic arch to reach the lower jaw; cheek muscles arise on the arch and go to the lower jaw.

A high level of metabolism requires mammals to eat a great deal of food and also to exchange a large volume of air in their lungs. Mammals must continue to breathe when they are feeding and chewing food. This is made possible by the evolution of a *hard palate,* which in life is continued caudally by a fleshy *soft palate.* Processes of the premaxillary, maxillary and palatine bones extend toward the midline and unite with each other ventral to the original palate. This displaces the openings of the internal nostrils posteriorly, thus permitting the animal simultaneously to breathe and manipulate food within its mouth. The anterior portion of the primitive palate, now situated dorsal to the secondary palate, regresses to some extent and this area is occupied by enlarged nasal cavities.

During the evolution of mammals, shifts occurred in the direction of pull of certain jaw muscles such that the resolution of the force of the muscle contraction made for a stronger bite force and less force was wasted at the jaw joint. This was accompanied by important changes in the jaws. The dentary bone of the lower jaw of early reptiles, on which the major jaw muscles attached, enlarged and the postdentary bones became smaller. As forces at the jaw joint decreased, the joint-bearing bones (quadrate and articular) could, and did, become much smaller. Eventually they were little more than a nubbin of bone lying posterior to a part of the dentary that had reached the squamosal. A new joint evolved between the dentary and squamosal, and the quadrate and articular disappeared from the jaw mechanism. The quadrate and articular were not lost, however, because they became associated with the auditory apparatus. There is evidence that a tympanic cavity and membrane were situated directly behind the jaw joint in mammal-like reptiles (see Fig. 5–6, p. 90), and some investigators have suggested that the quadrate and articular were part of a bone conducting system carrying sound waves from the lower jaw to the ear. The development of a new joint made it possible for certain of the primitive, posterior jaw elements to become specialized as parts of the auditory apparatus. The articular and quadrate have become additional auditory ossicles, the *malleus* and *incus,* respectively. The change in jaw joint and auditory ossicles is the osteological feature diagnostic of mammals.

The delicate auditory ossicles of mammals (malleus, incus, stapes) are protected by the formation of a plate of bone beneath the tympanic, or middle ear, cavity. Two elements contribute to this bony encasement in most instances—a cartilage replacement *endotympanic* having no homologue in lower vertebrates, and a dermal *tympanic,* homologous to the angular of the lower jaw of amphibians and reptiles.

Changes in dentition accompanied the changes in the jaws. The teeth of mammals are no longer

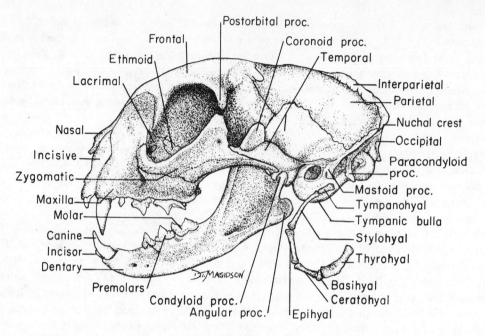

Figure 2-6   Lateral view of the skull, lower jaw, and hyoid apparatus of the cat.

simple cones adapted for seizing and holding prey; advantage is taken of the stronger bite, and the teeth are differentiated into types that can seize, cut, and crush the food.

(A) GENERAL FEATURES OF THE SKULL

Examine the skull of a cat (Figs. 2–6 and 2–7). As with other vertebrates, it can be divided into an anterior *facial region* containing the jaw, nose, and eyes, and a posterior

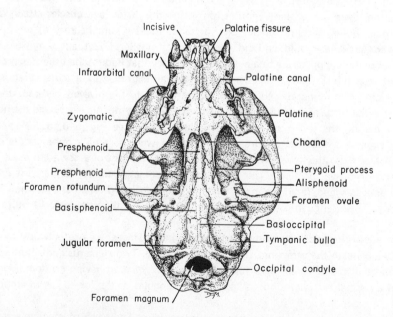

Figure 2-7   Ventral view of the skull of the cat.

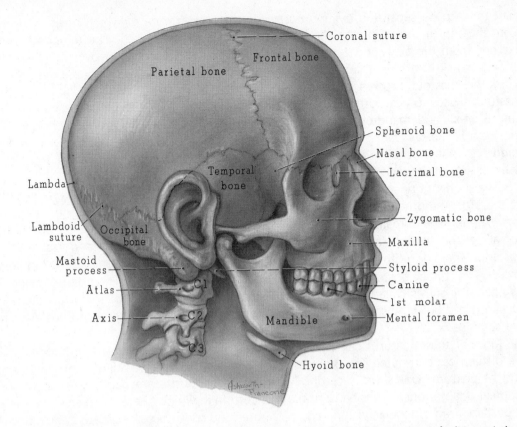

Figure 2–8 Lateral view of the skull, the lower jaw, and the hyoid apparatus of a human being. (From Jacob and Francone: Structure and Function in Man. W. B. Saunders Company.)

cranial region housing the brain and ear. Notice the external nostrils *(nares)* at the anterior end of the skull, and the large, circular *orbits* in which the eyeballs are lodged. Although the two nares may appear contiguous, they are separated in life by a fleshy and cartilaginous septum.

The *foramen magnum,* through which the spinal cord enters the skull, can be seen in the occipital region at the caudal end of the skull. There is an *occipital condyle* on each side of it. The large, round swelling on the ventral side, anterior to each occipital condyle, is the *tympanic bulla,* which encases the underside of the tympanic cavity. In life, the tympanum is lodged in the opening *(external acoustic meatus)* on the lateral surface of the bulla. The tongue-shaped bump of bone on the lateral surface of the bulla just posterior to the external auditory meatus is the *mastoid process.* A similarly shaped *paracondyloid process* is located on the posterior surface of the bulla. The human skull lacks the swollen tympanic bulla and has a much larger mastoid process (Fig. 2–8).

The handle-like bridge of bone on the side of the skull, extending from the front of the orbit to the external acoustic meatus, is the *zygomatic arch.* All of it is included within the facial portion of the skull. A *mandibular fossa* for the articulation of the lower jaw appears as a smooth groove on the ventral surface of the posterior portion of the arch.

A large *temporal fossa* is situated dorsal to the back half of the zygomatic arch and posterior to the orbit. Although temporal fenestra and orbit merge to some extent, they tend to be separated by two *postorbital processes,* one of which projects down from the top of the skull, the other up from the zygomatic arch. The temporal fossa extends from the orbit and zygomatic arch dorsally and posteriorly over much of the side of the skull. Its posterior border is the *nuchal crest,* a ridge of bone that extends from the mastoid process on one side, over the back and top of the skull, to the mastoid process on the other side. Cervical muscles attach on the caudal surface of the nuchal crest, and mandibular muscles on the cranial surface. The dorsal boundary of the fossa is the *sagittal crest,* a middorsal ridge of bone extending anteriorly from the middle of the nuchal crest, and a faint ridge *(temporal line)* that curves from the anterior end of the sagittal crest toward the dorsal postorbital process.

The shelf of bone that extends across the palatal region, from the teeth on one side to those on the other, is the *hard palate.* In life, it is continued posteriorly by a fleshy *soft palate.* The remains of the primitive primary palate of primitive tetrapods lie dorsal to the hard palate. Internal nostrils, or *choanae,* will be seen dorsal to the posterior margin of the hard palate. A nearly vertical plate of bone, the *pterygoid process,* continues posteriorly from each side of the hard palate. The point of bone at its posterior end is called the *hamulus.* A small *pterygoid fossa,* for the attachment of certain mandibular muscles, appears as an elongated groove just lateral and posterior to the hamulus.

Most of these features can also be seen in a human skull (Fig. 2–8), although there are differences correlated for the most part with our very large brain, short face, and upright posture. Our short face and forward-turned orbits are partly a result of the arboreal adaptations of our early primate ancestors in which the sense of smell was less important than binocular vision. The nasal area and snout became greatly reduced and this permitted the orbits to rotate forward and made possible an extensive overlap of the visual fields of the eyes. Orbit and temporal fossa are separated in human beings by a complete septum of bone. Since the skull is balanced on the top of the vertebral column, the foramen magnum is far under the skull rather than at the back as in a cat, and the nuchal crest on which certain neck muscles attach is less developed. Our large mastoid process forms a lever arm which increases the mechanical advantage of the sternocleidomastoid muscle in one of its actions, namely, preventing the skull from tipping too far backward.

(B) COMPOSITION OF THE SKULL

To better visualize the elements of the skull, one should examine a set of disarticulated bones along with the entire skull. Most of the top and lateral sides of the skull are formed by dermal bones (Figs. 2–6 and 2–8). A small, tooth-bearing *incisive* (premaxilla) is located ventral and lateral to each external nostril. It also contributes a small process to the hard palate. A *maxilla* completes the upper jaw. It, too, contributes a process to the anterior portion of the hard palate, sends one up anterior to the orbit, and forms the anteromost portion of the zygomatic arch. An incisive is present in human embryos, but it has fused with the maxilla by birth. A small, delicate *lacrimal* is located in the medial wall of the front of the orbit just posterior to the dorsal extension of the maxilla. It is often broken. The portion of the zygomatic arch ventral to the orbit is formed by the *zygomatic* (jugal). A large *temporal* forms the posterior portion of the arch and the adjacent cranial wall and encases the internal and middle ear.

The mammalian temporal bone has evolved through the fusion of a number of bones that are independent in lower vertebrates. Its zygomatic process, and that portion of it that helps to encase the brain *(squamous portion)*, represent the squamosal. Of course, the part encasing the brain is an inward extension of the original squamosal of the dermal roof. The mastoid process, and its inward extension into the cranial cavity, which will be seen in the sagittal section of the skull, represent a part of the chondrocranium (otic capsule) that encases the inner ear. This part of the temporal may be called the *petrosal portion.* The thick part of the tympanic bulla adjacent to the external acoustic meatus is the *tympanic* and is homologous with the dermal angular of the lower jaw of primitive tetrapods. Finally, a cartilage replacement *entotympanic* forms the rest of the bulla and completes the encasement of the tympanic cavity. It has no homologue in lower forms. The middle ear is encased in human beings by the tympanic and entotympanic, but they do not form a bulbous bulla.

You may be able to get glimpses of the auditory ossicles (a chain of three bones which transmit sound waves from the tympanum to the inner ear) by looking in the external acoustic meatus, but you should examine a special preparation to see them clearly. The *malleus* is roughly mallet-shaped. It has a long, narrow handle that attaches to the tympanic membrane and a rounded head that articulates with the incus. The *incus* is shaped like an anvil. It has a concave surface for the reception of the malleus, and two processes that extend from the main surface of the bone. One of these articulates with the head of the stirrup-shaped *stapes.* A pair of narrow columns of bone extend from the head of the stapes to a flat, oval-shaped foot plate which fits into the oval window of the otic capsule. A stapedial artery passes between the columns of the stapes embryologically, and also in the adult of some species.

Dermal bones along the top of the skull are a pair of small *nasals* dorsal to the external nostrils, a pair of large *frontals* dorsal and medial to the orbits, and a pair of large *parietals* posterior to the frontals. The parietals have a long suture with the squamous portion of the temporals. As was the case with the squamosal, much of each parietal and frontal represents flanges of the original roofing bones that grew down median to the temporal muscles and helped to cover the brain. The frontals are paired in human embryos but usually fuse together by the eighth year of life.

A large, median *occipital* surrounds the foramen magnum and forms the caudal surface of the skull. The paracondyloid processes and most of the nuchal crest are on this element. Ventrally, the occipital forms the floor of the braincase between the tympanic bullae. In the skulls of young cats, a triangularly shaped *interparietal* will be seen in the middorsal line in front of the occipital and between the posterior part of the two parietals. In older individuals, it often fuses with the occipital and parietals.

The occipital bone forms the posterior portion of the chondrocranium, and the otic capsule is incorporated in the temporal. The rest of the primitive chondrocranium is represented by a part of the sphenoid, the ethmoid, and the turbinate bones. Although a tiny portion of the *ethmoid* can sometimes be seen in the medial wall of the orbit posterior to the lacrimal, most of this bone, and the turbinates too, can be seen best in a sagittal section, described below.

The sphenoid can be observed in the entire skull (Fig. 2–7). Roughly, it forms the floor and part of the sides of the braincase anterior to the tympanic bullae and posterior to the hard palate. In the cat, it is divided into separate anterior presphenoid and posterior basisphenoid bones. These should be seen in a disarticulated skull to appreciate their extent. The *basisphenoid* includes the plate of bone on the underside of the skull just anterior to the ventral portion of the occipital; the posterior portion of the pterygoid process; the hamulus; the three posterior foramina of a row of four at the back of the orbit; and a wing-like process extending dorsally between the squamous portion of the temporal and the frontal. The midventral plate of bone has evolved from the chondrocranium. That portion of it which includes the three foramina and dorsal extension is called the alisphenoid, and it has evolved from the epipterygoid of primitive tetrapods. The epipterygoid, in turn, is a cartilage replacement bone derived from the first visceral arch of fishes (Fig. 2–5). There is some doubt as to the homologies of the pterygoid process and its hamulus, but the consensus is that this region includes several dermal bones located in the palate of more primitive vertebrates. The *presphenoid* includes a narrow midventral strip of bone lying between the bases of the pterygoid processes and lateral extensions that pass dorsal to the pterygoid processes to enter the medial wall of the orbits. The lateral portion contains the anterior foramen in the row of four referred to previously and has a common suture with the ventral extension of the frontal bone. The presphenoid has evolved from the chondrocranium. In some mammals, including human beings, presphenoid and basisphenoid are united and form a single bone.

Certain of the orginal dermal palatal bones are incorporated in the sphenoid; two others can be seen in more anterior parts of the skull. The *vomer* appears as a midventral strip of bone anterior to the sphenoid and dorsal to the hard palate. You will have to look into the internal nostrils to see it, for it helps to separate the nasal cavities. Paired *palatine* bones form the posterior portion of the hard palate, the anterior portion of the pterygoid processes, and a bit of the medial wall of the orbits posterior and ventral to the lacrimals.

(C) INTERIOR OF THE SKULL

Examine a pair of sagittal sections of the cat skull cut in such a way that the nasal septum shows on one half and the turbinates on the other. Note the large *cranial cavity* for the brain (Figs. 2–9 and 2–10). It can be divided into three parts—a *caudal cranial fossa* in the occipital-otic region for the cerebellum; a large *middle cranial fossa* anterior to this for the cerebrum; and a small, narrow *rostral cranial fossa,* just posterior to the nasal region, for the olfactory bulbs of the brain. In the cat, a partial transverse septum of bone (the *tentorium*) separates the middle and caudal cranial fossae. The internal part of the petrous portion of the temporal, containing the inner ear, can be seen in the lateral wall of the caudal cranial fossa dorsal to the tympanic bulla. It is perforated by a large foramen, the *internal acoustic meatus.* The small fossa dorsal to the internal acoustic meatus lodges a lobule of the cerebellum. The saddle-shaped notch *(sella turcica)* in the floor of the cerebral fossa lodges the hypophysis (Fig. 7–7, p. 129). It lies in the sphenoid. A large *sphenoidal air sinus* can be seen in the sphenoid region beneath the anterior portion of the cerebral fossa, and a *frontal air sinus* in the frontal bone dorsal to the rostral cranial fossa.

The anterior wall of the rostral cranial fossa is formed by a sievelike plate of bone whose foramina communicate with the nasal cavities. This plate of bone, the *cribriform*

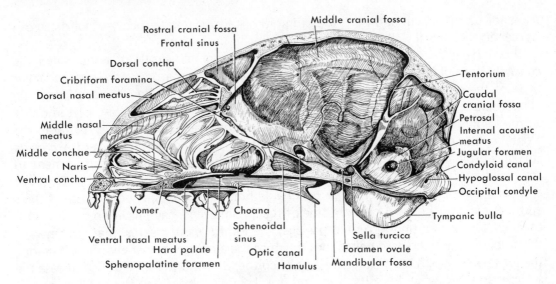

Figure 2–9 Sagittal section of the skull of a cat. The nasal septum has been removed.

plate of the *ethmoid,* can also be seen by looking through the foramen magnum of a complete skull. A *perpendicular plate* of the ethmoid, which will be seen on the larger section, extends from the cribriform plate anteriorly between the two nasal cavities. It connects with the nasal bones dorsally and with the vomer ventrally and forms much of the nasal septum. The very front of the septum is cartilaginous. Most of the ethmoid is shaped like the letter T, the top of the T being the cribriform plate, and its stem the

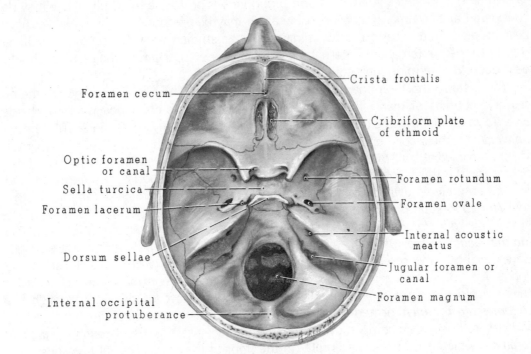

Figure 2–10 Floor of the human cranial cavity. (From Jacob and Francone: Structure and Function in Man. W.B. Saunders Company.)

perpendicular plate. In addition, the ethmoid has little lateral processes that often can be seen in the medial wall of the orbits posterior to the lacrimal (Fig. 2–6).

Each nasal cavity is largely filled with thin, complex scrolls of bone called the *conchae* (turbinates). They will show best in the section that lacks the perpendicular plate of the ethmoid (Fig. 2–9). Although the conchae ossify from the portion of the chondrocranium encapsulating the nose, they become attached to one of the bones surrounding the nasal cavities. A small *dorsal concha* (nasoturbinate) lies rostral to the frontal sinus and lateral to a perpendicular septum of the nasal bone. A *ventral concha* (maxilloturbinate) lies in the rostroventral part of the nasal cavity. Its attachment to the maxillary bone can be seen by looking into the cavity through the naris. The large and very complexly folded *middle concha* (ethmoturbinate) lies between the others and fills most of the nasal cavity. Some of the air entering a nasal cavity passes directly caudally through an uninterrupted passage, the *ventral nasal meatus,* that lies between the turbinates and the hard palate, but much of the air filters back between and among the turbinates, which are covered by a nasal epithelium, and which serve to increase the surface area available for olfaction and for heating, cleansing, and moistening the inspired air. A *dorsal nasal meatus* leads from the dorsal part of the naris caudally to the caudal parts of the middle concha. A narrow *common meatus* lies between all of the conchae and the nasal septum. The conchae of human beings are less complexly folded, correlating to a reduction in our sense of smell (Fig. 5–5, p. 89).

(D) FORAMINA OF THE SKULL

Numerous foramina for nerves and blood vessels perforate the skull. These may be considered at this time, or their consideration may be postponed until the nerves have been studied. In any case, the foramina should be studied on a section of the skull so that both their external and internal aspects can be seen. Certain of the passages should be probed to determine their course.

First, study the foramina for the 12 cranial nerves. The first cranial nerve, the olfactory, consists of many fine subdivisions that enter the cranial cavity through the *cribriform foramina* in the cribriform plate of the ethmoid (Figs. 2–9 and 2–10). The *optic canal,* for the second cranial nerve, is the most anterior of the row of four foramina in the posteromedial wall of the orbit. Next posterior, and largest in this row, is the *orbital fissure* for the third, fourth, and sixth nerves going to the muscles of the eyeball, and for the ophthalmic division of the fifth nerve. The third and smallest foramen in this row *(foramen rotundum)* transmits the maxillary division of the fifth nerve, and the last in this row *(foramen ovale),* the mandibular division of this nerve (Fig. 2–7). The *internal acoustic meatus,* seen in the petrosal when the sagittal section was studied, transmits both the seventh and eighth nerves. The eighth (vestibulocochlear) nerve comes from the inner ear, but the seventh (facial nerve) continues through a facial canal within the petrosal and emerges on the undersurface of the skull through the *stylomastoid foramen* located beneath the tip of the mastoid process. The ninth, tenth, and eleventh nerves, in company with the internal jugular vein, pass through the *jugular foramen,* which can be seen internally in the floor of the caudal cranial fossa posterior to the internal acoustic meatus. The jugular foramen opens on the ventral surface of the skull beside the posteromedial edge of the tympanic bulla. The *hypoglossal canal* for the twelfth cranial nerve can be seen in the floor of the caudal cranial fossa posterior and medial to the jugular foramen. Probe to see that the hypoglossal canal extends anteriorly and emerges on the ventral surface in common with the jugular foramen.

As the ophthalmic and maxillary divisions of the fifth (trigeminal) nerve are distributed to various parts of the head, certain of their branches pass through other foramina, often in company with blood vessels. A branch of the ophthalmic division traverses a small *ethmoid foramen,* or series of foramina, in the medial wall of the orbit to enter the nasal cavity. The ethmoid foramen lies in the frontal bone very near its suture with the sphenoid. One of the branches of the maxillary division emerges through an *infraorbital canal* located in the anterior part of the zygomatic arch, but before reaching the infraorbital canal, this branch has subsidiary branches that pass through several small foramina near the front of the orbit to supply the teeth of the upper jaw. A second branch of the maxillary division leaves the anterior portion of the orbit through a *sphenopalatine foramen* to enter the ventral part of the nasal cavity. The sphenopalatine foramen is the larger, and more medial, of two foramina that lie close together in the orbital process of the palatine bone. After entering the nasal cavity, part of this nerve continues forward through the nasal passages and finally drops down to the roof of the mouth through one of two *palatine fissures* located on each side of the midline at the anterior end of the hard palate. In the cat, but not in human beings, the palatine fissure also carries a nasopalatine duct that leads from the mouth to the vomeronasal organs (Jacobson's organs, p. 88) located in the nasal cavities. The small foramen lateral and anterior to the sphenopalatine is the posterior end of the *palatine canal.* A third branch of the maxillary division runs through this canal to the roof of the mouth. The anterior end of the canal can be seen on the ventral surface of the hard palate in, or near, the suture between the palatine and maxillary bones. A fourth branch of the maxillary backtracks to enter the orbital fissure. It then passes into a small *pterygoid canal* whose anterior end may be seen in the floor of the orbital fissure. The posterior end of the canal appears as a tiny hole on the ventral side of the skull between the ventral part of the basisphenoid and the base of the pterygoid process. After emerging from the pterygoid canal, this branch enters the tympanic cavity through the osseous portion of the *auditory tube*—a large opening at the anterior edge of the tympanic bulla. A summary of the cranial nerves and the foramina through which they pass is given in Table 2–2.

The major foramina that remain do not carry nerves. A *lacrimal canal,* for the nasolacrimal, or tear, duct will be seen in the lacrimal bone extending from the orbit into the nasal cavity. You may also be able to find parts of the *carotid canal* for the internal carotid artery. This artery is vestigial in the cat and the carotid canal is very small, but both are well developed in human beings. The posterior end of the canal appears as a tiny hole in the anteromedial wall of the jugular foramen. From here the canal extends forward, dorsal to the tympanic bulla, and enters the cranial cavity. Its point of entry can be seen in the caudal cranial fossa anterior to the petrosal and ventral to the tentorium. A *condyloid canal,* for a small vein, can be found in the caudal cranial fossa dorsal to the hypoglossal canal.

If you have access to a specimen in which the tympanic bulla has been removed, you will be able to see two openings on the underside of the petrosal. The more dorsal is the *fenestra vestibuli,* or oval window, for the stapes, the more ventral the *fenestra cochleae,* or round window, for the release of pressure waves from the liquid in the inner ear.

TABLE 2-2 *Mammalian Cranial Nerves and Their Foramina*

The twelve cranial nerves of mammals are listed together with the foramina through which they pass. In those cases in which a nerve goes through two or more foramina before reaching the organ it supplies, the foramina are listed in sequence from the brain.

Nerve	Foramen
I. Olfactory	Cribriform foramina
II. Optic	Optic canal
III. Oculomotor	Orbital fissure
IV. Trochlear	Orbital fissure
V. Trigeminal	
Ophthalmic division	Orbital fissure
one branch	Ethmoid foramina
Maxillary division	Foramen rotundum
one branch	Infraorbital canal
one branch	Sphenopalatine foramen, Palatine fissure
one branch	Palatine canal
one branch	Pterygoid canal
Mandibular division	Foramen ovale
one branch	Mandibular foramen, Mental foramina
VI. Abducens	Orbital fissure
VII. Facial	Internal acoustic meatus, Facial canal, Stylomastoid foramen
VIII. Vestibulocochlear	Internal acoustic meatus
IX. Glossopharyngeal	Jugular foramen
X. Vagus	Jugular foramen
XI. Accessory	Jugualr foramen
XII. Hypoglossal	Hypoglossal canal, Jugular foramen

(E) LOWER JAW

With the transfer of certain of the lower jaw bones of primitive tetrapods to the ear region, and the loss of others, the pair of enlarged dentaries are left as the sole elements in the mammalian lower jaw.

The lower jaw, or *mandible,* of mammals (Fig. 2-6) consists of a pair of *dentary bones* firmly united anteriorly at the *mandibular symphysis.* The horizontal part of the mandible that bears the teeth is its *body;* the part posterior to this, its *ramus.* A large, triangular-shaped depression, *masseteric fossa,* occupies most of the lateral surface of the ramus. Part of the masseter, one of the mandibular muscles, inserts here. Posteriorly, the ramus has three processes—a dorsal *coronoid process,* to which the temporal muscle attaches; a middle, rounded *condyloid process* for the articulation with the skull proper; and a ventral *angular process* to which the rest of the masseter and the pterygoid muscles attach.

A large *mandibular foramen* will be seen on the medial side of the ramus, and two small *mental foramina* on the lateral surface of the body near its anterior end. A branch of the mandibular division of the trigeminal nerve, supplying the teeth and the skin covering the lower jaw, enters the mandibular foramen and emerges through the mental foramina. Blood vessels accompany the nerve.

(F) TEETH

The teeth of mammals are quite different from those of lower vertebrates, for they are limited to the jaw margins, are set in deep sockets *(thecodont),* and are differentiated into various types *(heterodont).* Most adult mammals have in each side of each jaw a series of nipping incisors at the front, a large canine behind these, a series of cutting premolars behind the canine, and finally a series of chewing, or grinding, molars. The number of each kind of teeth present in a particular group of mammals may be expressed as a dental formula. For primitive placental mammals this was $\frac{3.1.4.3}{3.1.4.3} \times$ 2 = 44. Such an animal had three incisors, one canine, four premolars, and three molars in each side of each jaw. The number of teeth, and their structure, are adapted to the animal's mode of life; the molars especially being subject to much divergence among the groups of mammals.

Examine the teeth of the cat (Figs. 2–6 and 2–7). In each side of the upper jaw there are normally three *incisors* in the incisive bone followed by one *canine,* three *premolars,* and one very small *molar* in the maxillary bone. In the lower jaw there are three incisors, one canine, two premolars, and one large molar. The dental formula is therefore $\frac{3.1.3.1}{3.1.2.1}$.

During its evolution, the cat has lost the first premolar in the upper jaw, the first two in the lower, and all molars posterior to the first. The gap left between each canine and the premolars is called a *diastema.* The first of the remaining premolars, and the molar of the upper jaw, are more or less vestigial. But the last premolar of the upper jaw (phylogenetically premolar number four) and the lower molar have become large and complex in structure. Articulate the jaws and note how these two teeth, which are known as the *carnassials,* intersect to form a specialized shearing mechanism. Carnassials are restricted to carnivores, and in contemporary species have the formula Pm 4/M 1.

Correlated with the short facial region, we have fewer teeth than most mammals and the teeth are adapted for a diet of soft foods of many types. We are not specialized for a particular type of food as are many mammals. In each side of the upper and lower jaw we have two incisor teeth, one canine, two premolars (bicuspids), and three molars (Fig. 2–8), giving us a dental formula of $\frac{2.1.2.3}{2.1.2.3}$. The crown of the canine has come to resemble an incisor, but the root remains large. Our molars have square-shaped and flattened crowns well adapted for crushing food.

(G) HYOID APPARATUS

The extensive visceral skeleton, which supported the gills in fishes, is greatly reduced in terrestrial vertebrates, and especially so in mammals. You have already seen the fate of certain portions of it. Part of the first (mandibular) arch forms the alisphenoid and part the malleus and incus. The dorsal portion of the second (hyoid) arch gives rise to the stapes. Much of the rest of the visceral skeleton is simply lost, but the ventral portion of the hyoid arch and the third arch contribute to the hyoid apparatus, and parts of arches 4, 5 and 6 give rise to the laryngeal cartilages (p. 132).

The hyoid apparatus forms a bony sling that supports the base of the tongue (Fig. 2-6). It may be in place on the skeleton, or removed and mounted separately. It consists in the cat of a transverse bar of bone, the *body of the hyoid* or basihyals, from which two pairs of processes (horns) extend cranially and caudally. The caudal or *greater horns of the hyoid* are the larger, and each consists of but one bone, the *thyrohyal,* derived from the first branchial arch. Each of the cranial or *lesser horns of the hyoid* consists of a small *ceratohyal.* A chain of ossicles connects the ceratohyal with the skull, attaching to the tympanic bulla medial to the stylomastoid foramen. From ventral to dorsal these are the *epihyal, stylohyal,* and *tympanohyal.* These ossicles and the ceratohyal are derivatives of the hyoid arch. The body of the hyoid develops partly from the hyoid arch and partly from the first branchial arch.

In some mammals (human beings, for example Fig. 2–8) the hyoid apparatus consists of a single bone, the hyoid, having a body, a lesser horn, and a greater horn. A stylohyoid ligament, which extends from a styloid process at the base of the skull to the lesser horn, replaces the chain of ossicles seen in the cat as the support for the apparatus. The styloid process itself represents a part of the hyoid arch fused onto the temporal bone.

Chapter Three — THE APPENDICULAR SKELETON

As stated in the preceding chapter, the appendicular skeleton consists of the bones of the paired appendages, and the girdles to which the appendages attach. Most vertebrates have both *pectoral* (shoulder) and *pelvic* (hip) *girdles* and *appendages.* The appendicular skeleton is basically a part of the endoskeleton, so it consists primarily of cartilage or cartilage replacement bone. However, dermal bones have become intimately associated with the cartilaginous elements of the pectoral girdle in the majority of vertebrates.

EVOLUTION OF THE APPENDICULAR SKELETON

Fishes swim by lateral undulations of their trunk and tail; their paired appendages are fins, which are used primarily in stabilizing the body and in turning. During the transition from water to land, the appendages assumed primary importance in support and locomotion.

In primitive amphibians and reptiles, the limbs are in a sprawled position at right angles to the body and are used as a supplement to lateral undulations of the body in locomotion. In such a limb, the humerus and femur move back and forth close to the horizontal plane. In the evolution through more advanced reptiles to birds and mammals, the limbs become increasingly important in support and locomotion. In this connection the limbs of mammals have rotated nearly beneath the body so that the humerus and femur move back and forth close to the vertical plane. Review the changes in limb posture described in the section on External Anatomy (p. 5).

In correlation with the increased importance of the appendages, the girdles become larger and stronger. The dorsal portion of the pelvic girdle acquires an attachment to the sacral vertebrae and ribs, but weight is transferred from the front of the body to the pectoral girdle and appendage by a muscular sling (Fig. 4–11, p. 55). The limb posture of a primitive terrestrial vertebrate necessitates particularly powerful muscles on the ventral portion of the girdle; hence these parts of the girdle are relatively large in amphibians and reptiles. With the rotation of the limbs beneath the body and their fore and aft swing, dorsal musculature becomes more important, so we find that the dorsal part of the mammalian girdles is expanded and the ventral portions are somewhat reduced.

Our primitive primate ancestors were quadrupeds that scampered through the trees using all their limbs. The limbs remained very flexible, a grasping hand and foot evolved with an opposable thumb and great toe, and claws were transformed into finger and toe nails. Changes in locomotion occurred in the more advanced and larger primates; the apes became brachiators, swinging through the trees with their long arms, and man became an erect, terrestrial biped. Our hind legs are larger and stronger than our arms, and our feet are modified for supporting the entire body on the ground.

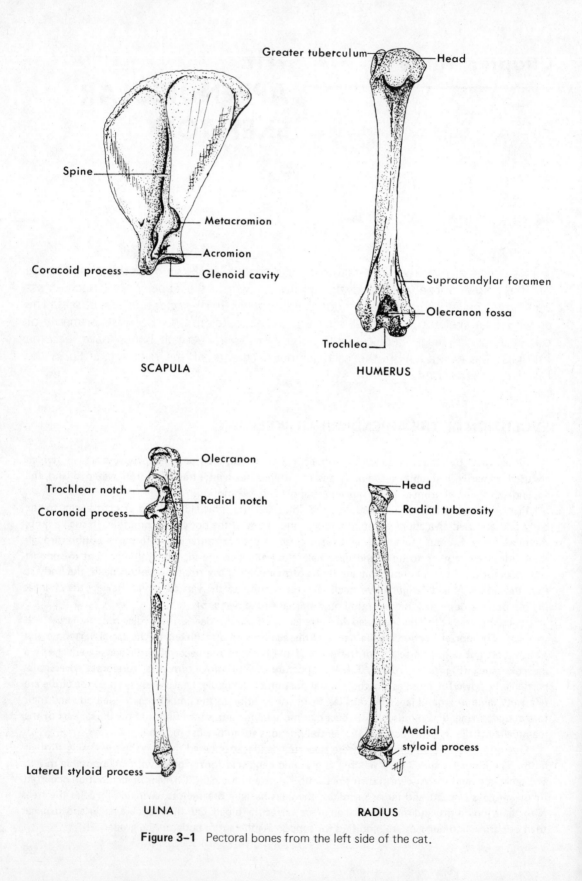

Figure 3–1   Pectoral bones from the left side of the cat.

PECTORAL GIRDLE AND APPENDAGE

Study the cat or human appendages and girdles on mounted specimens and from disarticulated bones (Figs. 3–1 and 3–2). Learn to distinguish the individual girdles and the long bones of the appendage, including a recognition as to whether they are from the left or right side. The pectoral girdle consists primarily of an expanded triangular-shaped *scapula.* In primitive terrestrial vertebrates, there is also a large coracoid plate that extends nearly across the ventral surface of the chest, but in most mammals this is reduced to a small, hooklike *coracoid process* which can be seen medial to the cranial edge of the *glenoid cavity* (socket for the articulation with the front leg). Of several dermal elements originally associated with the girdle, only the *clavicle* remains. In many mammals, ourselves included, the clavicle extends from the scapula to the sternum and helps to hold the shoulder joint in position. But in some mammals, including the cat, it loses its connections with the rest of the skeleton and is reduced to a sliver of bone embedded in the muscles cranial to the shoulder joint. This permits the scapula to participate in the fore and aft swing of the front leg during locomotion.

If one pictures the scapula as an inverted triangle, the glenoid cavity is at the apex, and the curved top of the scapula *(dorsal border)* is at the base of the triangle. The cranial edge of the scapula is its *cranial border* and the straight caudal edge, which is adjacent to the armpit (axilla), its *caudal border.* A prominent ridge of bone, the *scapular spine,* extends from the dorsal border nearly to the glenoid cavity. The ventral tip of the spine continues lateral to the glenoid cavity as a process known as the *acromion.* The clavicle articulates with this process in those mammals having a prominent clavicle. A *metacromion (hamate process)* extends caudally from the spine dorsal to the acromion; it is not prominent in human beings. That portion of the lateral surface of the scapula caudal to the spine is called the *infraspinous fossa;* that portion cranial to the spine, the *supraspinous fossa.* Since the spine represents the primitive cranial edge of the scapula, the portion of the scapula cranial to the spine is added during the course of evolution. The medial surface of the scapula is called the *subscapular fossa.* In human beings, from whom much of our anatomical terminology is derived, there is a prominent fossa here, but this surface is flat in the cat.

The bone of the brachium, or upper arm, is the *humerus.* Its expanded proximal end has a smooth rounded *head* that articulates with the glenoid cavity, and processes for muscular attachment—a lateral *greater tuberculum* and a medial *lesser tuberculum.* An *intertubercular groove* for the long tendon of the biceps muscle lies between the two tubercles on the craniomedial surface of the humerus. The distal end of the bone is also expanded and bears a smooth articular surface known as a *condyle.* It can be divided into a medial pulley-shaped portion (the *trochlea*) for the ulna of the forearm (the bone which comes up behind the elbow), and a lateral rounded portion (the *capitulum*) for the radius. You may have to articulate the ulna and radius with the humerus to determine the extent of trochlea and capitulum. As was the case with the subscapular fossa, these features are more obvious on a human humerus than on the cat's. An *olecranon fossa,* for the olecranon of the ulna, is situated proximal to the trochlea. The enlargements medial and lateral to the articular surfaces are the *medial epicondyle* and *lateral epicondyle,* respectively. A *supracondylar foramen* for the median nerve and radial artery is located above the medial epicondyle. This foramen is a primitive feature found in early reptiles but lost in most mammals, including human beings. That portion of the humerus, or of any long bone, lying between its extremities

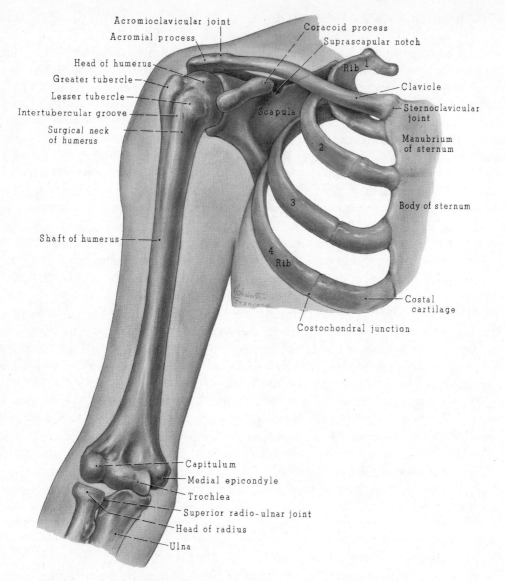

Acromioclavicular joint
Acromial process
Coracoid process
Suprascapular notch
Head of humerus
Greater tubercle
Rib 1
Lesser tubercle
Clavicle
Intertubercular groove
Scapula
Sternoclavicular joint
Surgical neck of humerus
2
Manubrium of sternum
3
Body of sternum
Shaft of humerus
4
Rib
Costal cartilage
Costochondral junction
Capitulum
Medial epicondyle
Trochlea
Superior radio-ulnar joint
Head of radius
Ulna

Figure 3–2 See legend on the opposite page.

is its *body* or shaft. The faint ridges and rugosites upon it mark the attachments of muscles.

The *ulna* is the longer of the two forearm bones, and a prominent *trochlear notch,* for the articulation with the humerus, will be seen near its proximal end. The end of the ulna lying proximal to the notch is the *olecranon* or "funny bone." A *coronoid process* forms the distal border of the notch, and a *radial notch,* for the head of the radius, merges with the trochlear notch lateral to the coronoid process. Distally the bone articulates with the wrist and bears a *lateral styloid process,* which is on the lateral side of the forearm when the hand is turned so that the palm is directed toward the ground. Note that the ulna and radius articulate distally and that the ulna plays a relatively insignificant role in the formation of the wrist joint. In some mammals, the distal half of the ulna is lost.

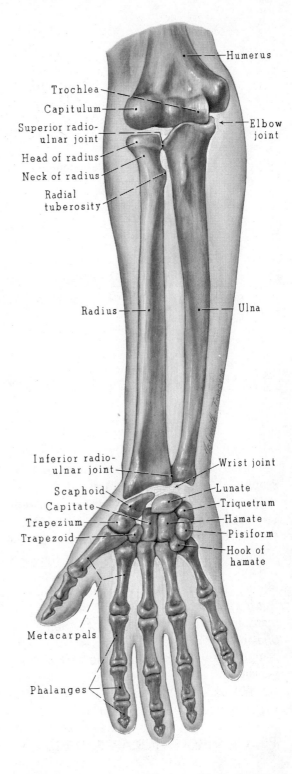

Figure 3–2 *Continued* Anterior view of the bones of the right pectoral girdle and arm of a human being. (From Jacob and Francone: Structure and Function in Man. W.B. Saunders Company.)

The other bone of the forearm is the *radius.* The articular surfaces on its *head* are of such a nature that the bone can rotate on the humerus and ulna. Slightly distal to the head is a prominent *radial tuberosity* for the insertion of the biceps muscle. The distal end of the radius is expanded, has articular surfaces for the ulna and carpus, and a short *medial syloid process.*

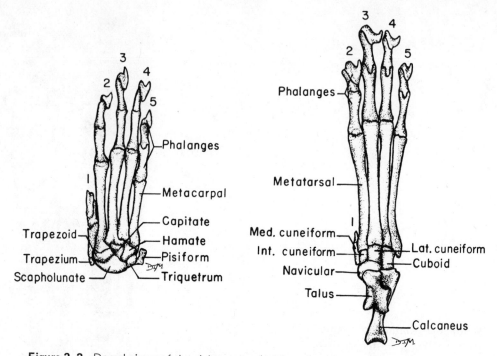

Figure 3–3 Dorsal views of the right manus (*left*) and pes (*right*) of the cat.

Study the hand *(manus).* Its first portion, the *carpus,* or wrist (Figs. 3–2 and 3–3), consists of two rows of small *carpal bones.* The proximal row contains a large medial *scaphoid,* which represents the radiale of more primitive tetrapods (Fig. 3–4); a *lunate,* which represents the intermedium; a *triquetrum,* which represents the ulnare and, on the lateral edge, a large, caudally projecting *pisiform.* Scaphoid and lunar are fused in the cat. The pisiform is one of many small sesamoid bones found in the appendages. They are not supporting elements; rather they are associated with the attachments of muscle tendons. Most are not named. The four elements of the distal row are, from medial to lateral, the *trapezium,* representing distal carpal 1; *trapezoid,* representing distal carpal 2; *capitate,* representing distal carpal 3; and *hamate,* representing distal carpal 4. Five *metacarpals* form the palm of the hand, and the free parts of the toes are composed of *phalanges.* The first toe is the most medial. The number of phalanges is less than in primitive tetrapods—two in the first digit and three in each of the others. The terminal phalanx of the catlike carnivores is articulated in such a way that it, and the claw which it bears, can be either extended or pulled back over the penultimate phalanx.

PELVIC GIRDLE AND APPENDAGE

The ilium, ischium, and pubis, which comprise the pelvic girdle of more primitive tetrapods, have fused together in adult mammals to form an *os coxae* (Figs. 3–5 and 3–6), but they can be seen clearly in young specimens. The *ilium* of the cat lies dorsal to the *acetabulum,* or socket for the hip joint. Its dorsalmost border is its *crest.* The ilium also inclines cranially and unites with more sacral vertebrae than does the ilium in

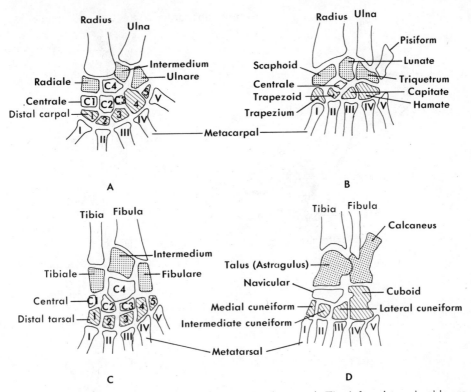

Figure 3–4 Evolution of the carpus (*top*) and talus (*bottom*). The left wrist and ankle are shown for a primitive amphibian (*A* and *C*) and for a mammal (*B* and *D*). (Modified after Romer: The Vertebrate Body. W.B. Saunders Company.)

primitive tetrapods. The *ischium* surrounds all but the cranial portion, and some of the medial side, of the large opening *(obturator foramen)* in the ventral portion of the girdle. Its enlarged caudolateral portion is referred to as the *tuberosity* of the ischium. The *pubis* lies cranial to the foramen and completes its medial wall. The pubes and ischia of opposite sides are united by *symphyses,* so that the pelvic girdle and sacrum form a complete ring, or *pelvic canal,* through which internal organs must pass to reach the anus and urogenital apertures. The pelvis of human beings contains the same bones, but there are differences in proportions. Our ilium is very broad and provides a large area for the attachment of the powerful gluteal muscles that help to hold the trunk in an upright position over the hind legs; only a pubic symphysis is present. Women have broader hips and larger pelvic canals than men, for the fetus must pass through this canal at birth.

Since the leg has rotated beneath the body, the thigh bone, or *femur,* articulates with the acetabulum by a *head* that projects from the medial side of the proximal end of the bone. A large, lateral *greater trochanter* and a small caudal *lesser trochanter* can also be seen on the proximal end. These processes are for muscle attachments. A depression called the *trochanteric fossa* is situated medial to the greater trochanter. The distal end of the femur has a smooth articular surface over which the *patella,* or kneecap (a sesamoid bone), glides. Posterior to this are smooth *lateral* and *medial condyles* for articulation with the tibia. The rough areas above each condyle are *epicondyles,* and the depression between the two condyles is the *intercondyloid fossa.*

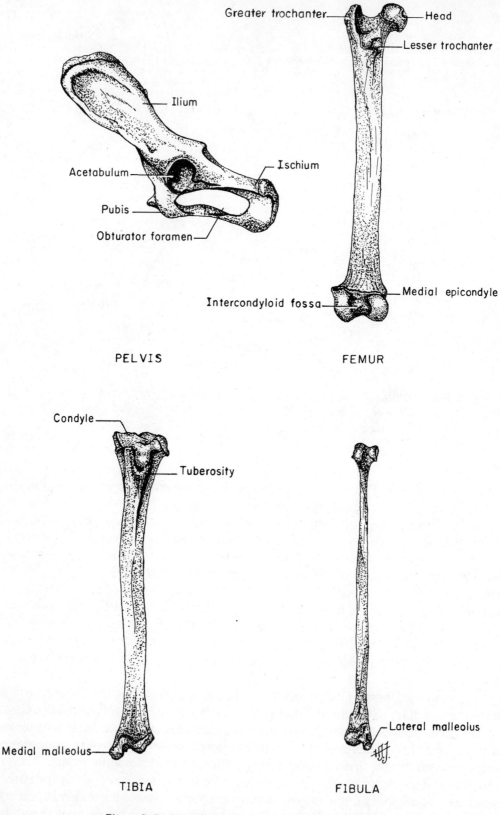

Figure 3–5 Pelvic bones from the left side of the cat.

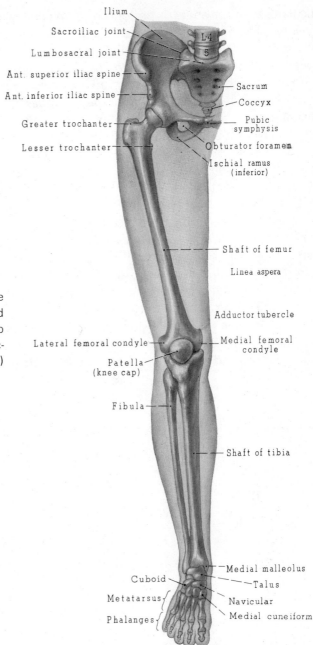

Ilium

Sacroiliac joint

Lumbosacral joint

Ant. superior iliac spine

Ant. inferior iliac spine

Greater trochanter

Lesser trochanter

L·4

5

Sacrum

Coccyx

Pubic symphysis

Obturator foramen

Ischial ramus (inferior)

Shaft of femur

Linea aspera

Adductor tubercle

Lateral femoral condyle

Patella (knee cap)

Fibula

Medial femoral condyle

Shaft of tibia

Cuboid

Metatarsus

Phalanges

Medial malleolus

Talus

Navicular

Medial cuneiform

Figure 3–6 Anterior view of the bones of the right pelvic girdle and leg of a human being. (From Jacob and Francone: Structure and Function in Man. W.B. Saunders Company.)

In a quadruped such as the cat the femur lies in a plane perpendicular to the ground, but in human beings it has an inward slant so that its distal end lies more medial than its proximal end (Fig. 3–6). This is a bipedal adaptation that brings the foot nearly over the projection of the body's center of gravity so that we can raise one foot from the ground without losing our balance.

The *tibia* is the larger and more medial of the two shank bones. Its proximal end has a pair of *condyles* for articulation with the femur, and an anterior, oblong bump (the *tuberosity*) for the attachment of the patellar ligament. Its shaft has a prominent anterior *crest* which continues from the tuberosity. The distal end of the tibia, which articulates with the ankle, is prolonged on the medial side as a process called the *medial malleolus.* The *fibula* is a very slender bone. Notice that it does not enter the knee joint but does serve to strengthen the ankle laterally. Its distal end has a small pulley-like process known as the *lateral malleolous.* Tendons pass posterior to this process.

Examine the foot *(pes).* The ankle, or *tarsus,* of the cat is typical of that of mammals, for it consists of seven *tarsal bones* (Figs. 3–3 and 3–6). The one that articulates with the tibia and fibula is the *talus* (astragalus), which appears to be homologous to the tibiale, intermedium, and one centrale of more primitive vertebrates (Fig. 3–4). The main joint in the mammal ankle is between this bone and the leg, rather than in the middle of the tarsus as in certain reptiles and birds. The large, posteriorly projecting heel bone is called the *calcaneous* (fibulare). A *navicular* (centrale) lies just distal to the talus, and a row of four bones lies distal to the navicular and calcaneus. There are, from medial to lateral, the *medial cuneiform* (distal tarsal 1), *intermediate cuneiform* (distal tarsal 2), *lateral cuneiform* (distal tarsal 3), and *cuboid* (distal tarsal 4). Five elongated *metatarsals,* which occupy the sole of the foot, normally follow the tarsals, and the free part of the digits is composed of *phalanges.* The cat has a digitigrade foot posture, and the first toe has been lost except for its metatarsal, which has been reduced to a small nubbin of bone. In contrast our first toe is the largest, for the medial side of the foot is close to the projection of the body's center of gravity and bears more weight than the lateral side. The first toe is in line with the others and has lost the grasping ability present in arboreal primates.

Chapter Four

THE MUSCULAR SYSTEM

GROUPS OF MUSCLES

Continuing on the general theme of the organ systems whose activities support and move the body, we will next consider the muscular system. In order to understand their evolution, the numerous muscles of vertebrates must be grouped in some way. Although the subdivisions of the muscular system according to histological structure (smooth vs. striated), or general type of innervation (involuntary vs. voluntary), are useful for certain types of work, the phylogenetically most natural method of subdivision is according to their mode of embryonic development in lower vertebrates (Fig. 4–1), for this sorts the muscles into groups which have a common evolutionary

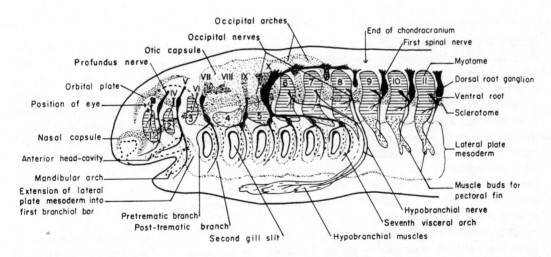

Figure 4–1 Diagram of the head of an embryo dogfish *(Scyllium)* to show the segmentation of the head and the origin of the cranial muscles. The embryo of a mammal resembes this in its major respects. The myotomes, which give rise to the somatic muscles, are horizontally hatched and numbered with Arabic numerals. The lateral plate mesoderm, which gives rise to the visceral musculature, is diagonally hatched. This part of the mesoderm contains the coelom and sends processes up between the gill slits into each of the interbranchial septa. The positions of the visceral arches are shown by broken lines. The nerves are shown in solid black and numbered with Roman numerals. The heavy stippling dorsally represents cartilage of the chondrocranium and vertebrae. (Slightly modifed after Goodrich: On the development of the segments of the head of Scyllium. Quarterly Journal of Microscopical Science, Vol. 63.)

TABLE 4–1 Groups of Vertebrate Muscles

Those groups of muscles innervated by cranial nerves are indicated; others are supplied by spinal nerves.

Somatic Muscles (From embryonic myotomes; somatic motor nerves)
    Axial muscles
        Extrinsic Ocular Muscles (Oculomotor, trochlear, and abducens nerves)
        Epibranchial Muscles
        Hypobranchial Muscles (Hypoglossal and spinal nerves)
           Prehyoid Muscles
           Posthyoid Muscles
        Trunk and Tail Muscles
           Epaxial Muscles
           Hypaxial Muscles
    Appendicular Muscles
        Dorsal Appendicular Muscles
        Ventral Appendicular Muscles
Visceral Muscles (From deep part of embryonic lateral plate; visceral motor nerves)
    Branchiomeric Muscles
        Mandibular Muscles (Trigeminal nerve)
        Hyoid Muscles (Facial nerve)
        Muscles of Third Visceral Arch (Glossopharyngeal nerve)
        Muscles of Remaining Arches (Vagus and accessory nerves)
      Muscles of Gut and Associated Structures (Vagus and spinal nerves)

history. Using this basis, there are two major groups of muscles—the somatic and the visceral (Table 4–1). *Somatic muscles* (parietal muscles) are associated with the body wall and appendages, and they develop from a long series of embryonic myotomes, one pair per body segment. They comprise the segmental trunk and tail muscles of fish, which obviously play an important role in swimming; a few somatic muscles enter the fins. Mammal embryos still have myotomes, but trunk muscles derived from them constitute less of the body mass in a mammal than in a fish, and most of the trunk muscles lose their segmentation. Appendicular muscles, in contrast, are much more complex and better developed. *Visceral muscles* are more deeply situated, being associated with the digestive tract. They develop from the deeper layer of the so-called "lateral plate mesoderm" (Fig. 4–1), and constitute the muscles of the gut, including the wall of the pharynx and its visceral arches. Although we generally think of these muscles as being smooth and involuntary, those of the visceral arches are striated and to a large extent voluntary.

Both the somatic and visceral muscles can be further subdivided. The somatic muscles are broken down into axial and appendicular groups. The *axial muscles,* as the name implies, are the muscles in the axis of the body—the extrinsic muscles of the eyeball, the epibranchial and hypobranchial muscles which in fish lie, respectively, dorsal and ventral to the gill region, and the muscles of the trunk and tail. In jawed vertebrates the hypobranchial and trunk muscles are further subdivided. The hypobranchial musculature is divided at the level of the hyoid into prehyoid and posthyoid groups; the trunk musculature into epaxial muscles lying dorsal and lateral to the vertebral column, and hypaxial muscles more ventrally. All these muscles are supplied by somatic motor fibers (p. 95) in spinal nerves and cranial nerves III, IV, VI, and XII.

The *appendicular muscles* develop embryologically from mesenchyme that lies within the limb bud. In the lower vertebrates, at least, this mesenchyme is derived early in development from myotomic buds. Thus, the appendicular muscles are closely related to axial muscles but are sufficiently distinct to warrant separate treatment. Since the mesenchyme in the limb bud becomes divided into dorsal and ventral premuscular masses before giving rise to the muscles, the appendicular muscles may be further subdivided into dorsal and ventral groups. True appendicular muscles lie within the appendage or on the girdle or may grow from the appendicular skeleton back onto the trunk. They should not be confused with muscles that start their development in some other part of the body and then grow over to the girdle. Appendicular muscles, too, are supplied by spinal nerves.

The visceral muscles are subdivided into two categories: (1) the *branchiomeric muscles* primitively associated with the visceral skeleton, and (2) the remaining muscles of the gut tube and associated structures. Only the branchiomeric group of visceral muscles will be considered. They, in turn, are grouped according to the arch with which they become associated—mandibular muscles, hyoid muscles, muscles of the third arch, etc. Branchiomeric muscles are innervated by visceral motor fibers in cranial nerves associated with the various visceral arches. The trigeminal nerve (V) innervates the muscles of the first arch; the facial (VII) supplies the muscles of the second arch; the glossopharyngeal (IX), the third; and the vagus (X) and accessory (XI) supply the musculature of the remaining arches.

MUSCLE TERMINOLOGY

In describing the individual muscles, a number of terms will be used, and these may appropriately be defined at this time. The ends of muscles are attached to skeletal elements, or to connective tissue septa, and the intermediate part of the muscle (its belly) is free. One end of a muscle tends to be fairly stationary during contraction, and the other end moves the bone to which it is attached as the muscle shortens. The more stable end is the *origin* of the muscle; the opposite end is the *insertion* (Fig. 4-2A). In the case of limb muscles, the origin is proximal and the insertion distal. A muscle may have two or more points of origin, or two or more points of insertion. If the multiple points of attachment are segmentally arranged, they are called *slips.* The multiple origins of a muscle which are not segmented are sometimes called *heads.*

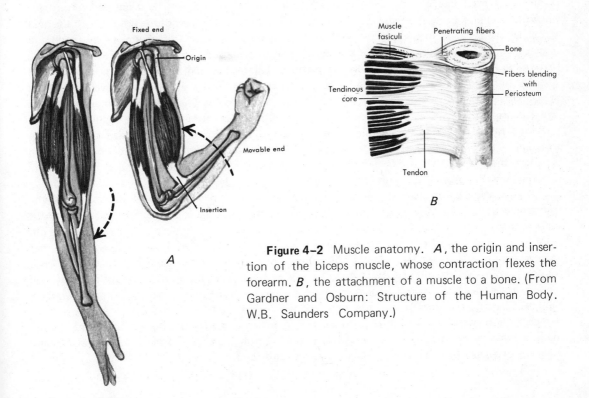

Figure 4-2 Muscle anatomy. *A*, the origin and insertion of the biceps muscle, whose contraction flexes the forearm. *B*, the attachment of a muscle to a bone. (From Gardner and Osburn: Structure of the Human Body. W.B. Saunders Company.)

Connective tissue forms a covering about a muscle and spreads into the muscle, where it surrounds bundles of muscle fibers (muscle fascicles) and even invests the individual muscle cells. The connective tissue covering of a muscle is a part of the *deep fascia*—a dense layer of connective tissue that forms a sheath for the individual muscles and also may hold groups of muscles together. Deep fascia is to be distinguished from *superficial fascia,* which is the layer of loose connective tissue beneath the skin. Superficial fascia generally contains fat; deep fascia does not.

The attachment of a muscle to a bone is not by the actual muscle fibers; rather it is made by the connective tissue that surrounds groups of muscle fibers and invests the entire muscle (Fig. 4–2, *B*). This connective tissue merges with the periosteum covering the bone, and some of the connective tissue fibers may penetrate the bone. If the muscle fibers come very close to the bone, and the muscle appears to attach to it, we speak of a "fleshy" attachment; if a narrow band of connective tissue extends from the muscle to the bone, we speak of a *tendon;* if the muscle attaches by a broad, thin sheet of connective tissue, we speak of an *aponeurosis.*

Sometime a transverse septum of connective tissue, called a *tendinous intersection,* is found in the middle of a muscle. Such an intersection often represents either a persistent myoseptum or the point at which two originally separate muscles have united.

ACTION OF MUSCLES

Most muscles act in antagonistic groups; that is, the action of one muscle, or group of muscles, is offset by an antagonistic muscle or group. Various sets of terms describe these antagonistic actions. In quadrupeds, *protraction* defines the forward movement of the humerus, the femur, or the entire limb in the longitudinal plane; *retraction* is the backward movement. In human beings, these actions are sometimes described as flexion and extension, respectively. *Adduction* and *abduction* define the movement of a part toward or away from a point of reference respectively. For a limb, the reference is the midventral line of the body; thus adduction of the arm moves its distal end toward the midventral line of the body. *Flexion* is the movement of a distal segment of the limb toward the next proximal segment, as in the approximation of the antebrachium and brachium, or the hand and antebrachium. Flexion also describes the bending of the head or trunk toward the ventral surface. *Extension* is movement in the opposite directions. Lateral bending of the trunk is called lateral flexion. In describing cat muscles, the terms flexion and extension are limited to movements of the distal parts of the appendage and certain movements of the trunk. The terms flexion and extension are also used in human anatomy to describe movements of the entire limb at the shoulder and hip. *Rotation* can be illustrated by the movement of the radius on the ulna, or the axis on the atlas. In rotation of the radius on the ulna, special terms are often used. Rotation of the forearm to a position in which the palm of the hand faces ventrally, or toward the ground, is called *pronation;* the opposite rotation, which brings the palm up, is *supination.*

The muscle actions described in this manual are the actions that would be caused by the shortening of the muscle in question acting alone. Such knowledge is a necessary first step toward understanding muscle action, but it should be recognized that it is an oversimplification. Electromyographic studies, which measure the electrical discharge of active muscles, show that muscle action is exceedingly complex. All parts of a muscle do not necessarily contract when a muscle becomes active, and seldom does one muscle contract independently of others. The synergistic action of a neighboring muscle frequently alters the force or direction of pull of the primary acting muscle. The degree of force to be overcome, the posture of the body, and the angle of the joint at which motion occurs may all affect the pattern of muscle activity. During flexion of the fingers, caused primarily by certain forearm muscles, muscles as distant as those of the shoulder may become active, presumably to stabilize the shoulder and the upper arm.

NAMES OF MUSCLES

Muscle names often are confusing to a student, but they can be very helpful, for they describe one or more distinguishing features of a muscle: its shape (trapezius); location (temporalis); shape and location (rectus abdominis); location and fiber direction (external oblique); primary action (tensor fasciae latae); or attachments (coracobrachialis).

THE STUDY OF MUSCLES

As methods of body support changed during the transition from fish to terrestrial vertebrate, and movements of the body and its parts became more complex during the evolution of vertebrates, the muscles became more numerous. It is not possible, in a course of this scope, to study all of them. Details of the muscles confined to the hand and foot have been omitted, as have those of the tail and perineum, but other groups are described with reasonable completeness. Insofar as possbile, the muscles are described by natural groups. This will permit a further selection by the instructor of the muscles to be studied if time is short.

A few remarks concerning the dissection of muscles may be appropriate. Insofar as possible confine your dissection of the muscles to one side of the body and cut open the body, when you do, on the opposite side. After the skin has been removed, the muscles must be carefully separated from each other. This involves cleaning off the overlying connective tissue with forceps until you can see the direction of the muscle fibers. Ordinarily, the fibers of one muscle are held together by a sheath of connective tissue, and all run in the same general direction to a common tendon or attachment. The fibers of an adjacent muscle will be bound together by a different sheath and will have a different direction and attachment. This will give you a clue as to where to separate one from the other. Separate the muscles by picking away, or tearing, the connective tissue between them with forceps, watching the fiber direction as you do so. Do not try to cut muscles apart. If the muscles separate as units, you are doing it correctly; but if you are exposing small bundles of muscle fibers, you are probably separating the parts of a single muscle.

It is best to expose and separate a few muscles of a given region before attempting to identify them. It will be necessary sometimes to cut through a superficial muscle to expose deeper ones. In such cases, you usually should cut across the belly of a muscle at right angles to its fibers, and turn back (reflect) its ends, rather than detach its origin or insertion. The dissection will be more meaningful if you have a skeleton before you on which to visualize the points of attachment of the muscles.

SKINNING AND CUTANEOUS MUSCLES

Lay your specimen on its belly, and make a middorsal incision through the skin that extends from the back of the head to the base of the tail. Make additional incisions from this cut around the neck; around the tail, anus, and external genitals; down the lateral surface of each leg; and around the wrists and ankles. Skin is to be left for the time being on the head, tail and perineum, and feet.

Beginning on the back, gradually separate the skin from the underlying muscles by tearing through the superficial fascia with a pair of blunt forceps. As you separate the skin from the trunk, notice the fine, parallel, brown lines that adhere to its undersurface. They represent the *cutaneous trunci,* the largest of several integumentary muscles that move the skin. The cutaneous trunci arises from the surface of certain appendicular muscles of the shoulder (latissimus dorsi and pectoralis), and from a midventral band of connective tissue *(linea alba),* fans out over most of the trunk, and

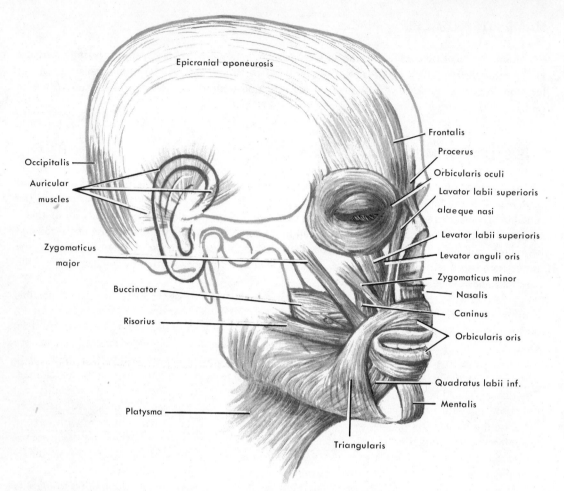

Epicranial aponeurosis

Frontalis

Procerus

Orbicularis oculi

Lavator labii superioris alaeque nasi

Levator labii superioris

Levator anguli oris

Zygomaticus minor

Nasalis

Caninus

Orbicularis oris

Quadratus labii inf.

Mentalis

Occipitalis

Auricular muscles

Zygomaticus major

Buccinator

Risorius

Platysma

Triangularis

Figure 4–3 Lateral view of the human platysma and facial muscles. (From Gardner and Osburn: Structure and Function of the Human Body. W.B. Saunders Company.)

inserts on the underside of the skin. It should be removed with the skin, except for that portion attached to the shoulder muscles caudal to the armpit. Several smaller cutaneous muscles become associated with the caudal part of the cutaneous trunci. They may not be noticed. The cutaneous trunci shakes the skin and helps the animal get rid of external foreign objects; it is not present in human beings.

Much of the top and sides of the neck is covered by another cutaneous muscle, the *platysma,* which is derived from the hyoid musculature. As may be noticed later when the head is skinned, the platysma, as it spreads over the face, breaks into a number of smaller muscles associated with the lips, nose, eyes, ears, etc. They are collectively known as the *facial muscles.* Platysma and facial muscles are well developed in human beings (Fig. 4–3). They are used for our numerous facial expressions.

As you continue to skin your specimen, you will come upon narrow tough cords passing to the skin. These are *cutaneous blood vessels* and *nerves* and must be cut. Note that they tend to be segmentally arranged along the trunk. If your specimen is a pregnant or lactating female, the *mammary glands* will appear as a pair of large, longitudinal, glandular masses along the ventral side of the abdomen and thorax. They should be removed with the skin.

After the specimen is skinned, clean away the excess fat and superficial fascia on the side that is to be studied, but do not clean an area thoroughly until it is being studied. If your specimen is a male, be particularly careful in removing the wad of fat in the groin, for it contains on each side the proximal part of the *cremasteric pouch*—a part of the scrotum containing blood vessels and the sperm duct extending between the abdomen and scrotal skin. First find this pouch. It is rather narrow. Clean away connective tissue deep in the groin, or inguinal region, and locate the boundary between the thigh and abdomen.

CAUDAL TRUNK MUSCLES

All the axial muscles cannot be studied at the same time, for many of them are located deep to the shoulder muscles. Those located on the trunk between the pectoral and pelvic appendages will be examined now, and the more cranial ones will be considered after the appendages have been studied (p. 73).

(A) HYPAXIAL MUSCLES

Continue to clean off the surface of the trunk between the pectoral and pelvic appendages. The wide sheet of tough, white fascia covering the lumbar region on the back is the *thoracolumbar fascia.* The wide sheet of muscle that runs cranially and ventrally from the cranial part of this fascia and disappears in the armpit is the latissimus dorsi (an appendicular muscle) (Fig. 4–6, p. 49). The large triangular muscle that covers the underside of the chest is the pectoralis (another appendicular muscle). The borders of the latissimus dorsi and pectoralis appear to run together caudal to the armpit. Separate the two in this region by removing the cutaneous trunci and carefully trace their edges forward. Lift up the caudoventral edge of the latissimus dorsi and the caudolateral edge of the pectoralis, and remove the fat and loose connective tissue from beneath them. The hypaxial trunk muscles can now be studied.

As in other tetrapods, the abdominal wall of a cat or human being is composed of three layers of muscle, plus a paired longitudinal muscle along the midventral line. All serve to compress the abdomen. The *external oblique* forms the outermost layer (Figs. 4–4 and 4–5). This muscle arises by slips from the surface of a number of caudal ribs and from the thoracolumbar fascia. Part of its origin lies beneath the caudal edge of the latissimus dorsi. Its fibers then extend obliquely caudally and ventrally to insert by an aponeurosis along the length of the linea alba. Using a sharp scalpel, or a razor blade, make a cut 2 or 3 inches long through the dorsolateral part of the external oblique at right angles to the direction of its fibers. Do not cut deeply, and as you cut watch for a deeper layer of muscle having a different fiber direction. When you reach the deeper layer, reflect part of the external oblique.

The *internal oblique* lies beneath the external oblique. It is most apparent high up on the side of the abdomen near its main origin from the thoracolumbar fascia. Its fibers extend ventrally and slightly cranially at right angles to the fibers of the external oblique and soon lead into a wide aponeurosis that inserts along the linea alba. Only the dorsal half of the muscle is fleshy. The ventral half is represented by an aponeurosis, but this may not be apparent at first, for one can see the third muscle layer, the transversus abdominis, through this aponeurosis.

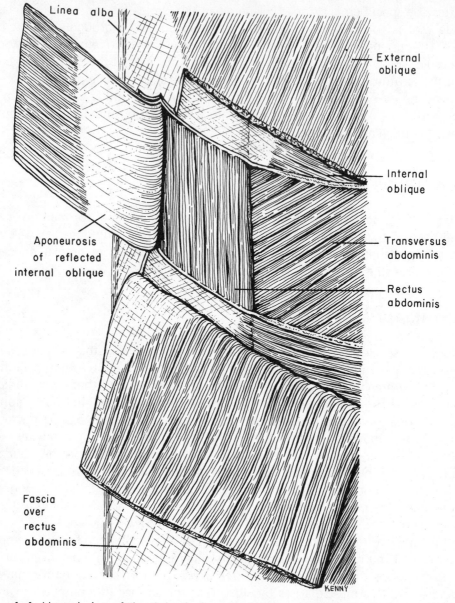

Figure 4–4 Ventral view of the abdominal muscle layers on the left side of the cat. Part of the external oblique and internal oblique have been reflected.

To get at the *transversus abdominis,* make a longitudinal cut through the fleshy portion of the internal oblique and reflect a part of this muscle. The transversus abdominis arises primarily from the medial surface of the more caudal ribs and from the transverse processes of the lumbar vertebrae. The latter portion of the origin lies in a furrow deep to the epaxial muscles. The fibers of the muscle extend ventrally, and slightly caudally, to insert along the linea alba by a narrow aponeurosis. Separate some of the fibers of the transversus, and you will expose the *parietal peritoneum* lining the abdominal cavity.

The reflection of the internal oblique also exposes a longitudinal band of muscle lying lateral to the midventral line. This is the *rectus abdominis.* It arises from the

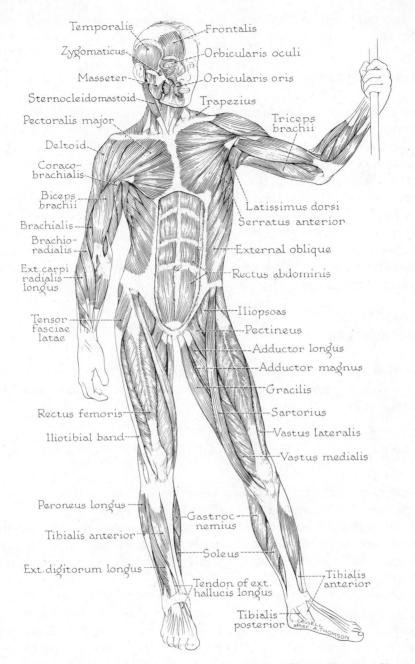

Figure 4–5 Anterior view of the muscles of a human being. (From King and Showers: Human Anatomy and Physiology. W.B. Saunders Company.)

pubis and passes cranially to insert in the cat on the cranial costal cartilages and sternum; it does not extend so far cranially in human beings. For much of its course, it lies between the aponeurosis of the internal oblique and the transversus. Transverse tendinous intersections can sometimes be seen in it.

In addition to the muscular layers of the abdominal wall, the caudal hypaxial musculature includes a subvertebral group that lies ventral to the lumbar and caudal thoracic vertebrae. This group, which includes the *quadratus lumborum* and *psoas minor,* is associated with certain pelvic muscles and is described in connection with the hind leg (p. 67).

(B) EPAXIAL MUSCLES

Lift up the thoracolumbar fascia with a pair of forceps, and make a longitudinal incision through it about one centimeter to one side of the middorsal line. Extend the incision from the latissimus dorsi to the sacral region and reflect the superficial layer of the fascia. A deeper layer of this fascia will now be seen encasing the epaxial muscles. Make a longitudinal cut through it about one centimeter from the middorsal line. The narrow band of muscle beside the spinous processes of the vertebrae is the *multifidus* (Fig. 4–24, p. 75). The wider lateral band of muscle represents the *erector spinae.* In the cat, a part of the fascia covering the erector spinae dips into, and subdivides, the erector spinae, but these subdivisions do not correspond with the divisions of the erector spinae on the cranial part of the trunk. Ignore them until the cranial part of the epaxial musculature is studied.

PECTORAL MUSCLES

Most of the muscles of the pectoral region are, of course, appendicular muscles, but a number of axial and several branchiomeric muscles have become associated with the girdle and appendage.

(A) PECTORALIS GROUP

The large triangular mass of muscle covering the chest is the *pectoralis.* It arises from the sternum and passes to insert primarily along the humerus. Its major actions are to pull the humerus toward the chest (adduction) and caudally (retraction), but in quadrupeds it also helps in transferring body weight from the trunk to the pectoral girdle and appendage (p. 54). Clean the surface of the muscle enough so that you can see the direction of its fibers.

A cleidobrachialis muscle covers the front of the shoulder cranial to the pectoral complex (Fig. 4–6). It will be considered later, but it should be mobilized at this time because part of the pectoralis is located deep to it. Lift up the cleidobrachialis and clean away connective tissue from beneath it. The clavicle is imbedded on the underside of the medial portion of this muscle. Cut through the connective tissue that binds the clavicle to the manubrium so that you can push the cleidobrachialis forward.

The pectoralis of quadrupeds is divided into a pectoralis superficialis and a pectoralis profundus. These two parts are homologous to the pectoralis major and the pectoralis minor of human beings, respectively, but the profundus is the larger muscle in quadrupeds. Each part may be further subdivided.

The *pectoralis superficialis* arises from approximately the cranial one third of the sternum. Its fibers extend more or less laterally to insert on the humerus. The insertion in the cat extends to the distal end of the humerus and onto the antebrachium. In the cat, the superficial pectoral can be divided into two parts. A narrow band of very superficial fibers, the *pectoralis descendens,* extends from the front end of the sternum

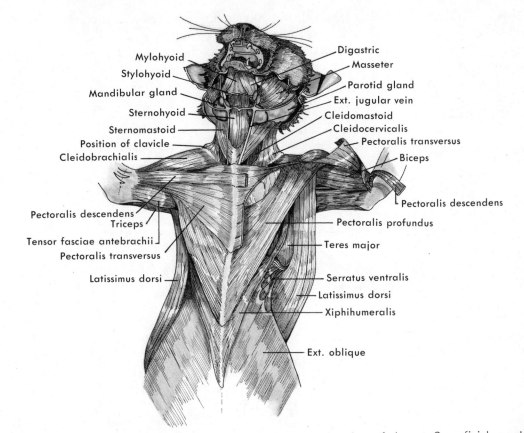

Figure 4–6 Ventral view of the muscles in the pectoral region of the cat. Superficial muscles are shown on the left side of the drawing; deeper muscles, after removal of the pectoralis transversus, on the right side.

to the antebrachium. The rest of the pectoralis superficialis is known as the *pectoralis transversus* (Fig. 4–6).

The *pectoralis profundus* originates from approximately the caudal two thirds of the sternum. Its fibers extend craniolaterally, disappear beneath those of the superficial pectoral, and insert on the humerus. The most caudal fibers of the cat's pectoralis profundus form a distinct, narrow band known as the *xiphihumeralis.*

Three muscles comprise the pectoral complex of human beings. A large, fan-shaped pectoralis major occupies the ventral surface of the chest and inserts upon the humerus (Fig. 4–5); a smaller pectoralis minor lies deep to it and inserts on the coracoid process of the pectoral girdle; and a very small subclavius lies deep to the minor and extends from the first rib to the clavicle.

(B) TRAPEZIUS AND STERNOCLEIDOMASTOID GROUP

The muscles belonging to the trapezius and sternocleidomastoid group are branchiomeric muscles that have become associated with the pectoral girdle, for they evolved from a gill arch muscle of fishes. The trapezius is a single, triangular-shaped muscle in human beings (Fig. 4–7), located on the back of the shoulder, but it is subdivided in the cat. The muscle and its parts act upon the girdle, primarily the scapula. They pull it toward the middorsal line (abduction), cranially (protraction) and caudally (retraction). The most cranial part of the trapezius (cleidocervicalis) inserts

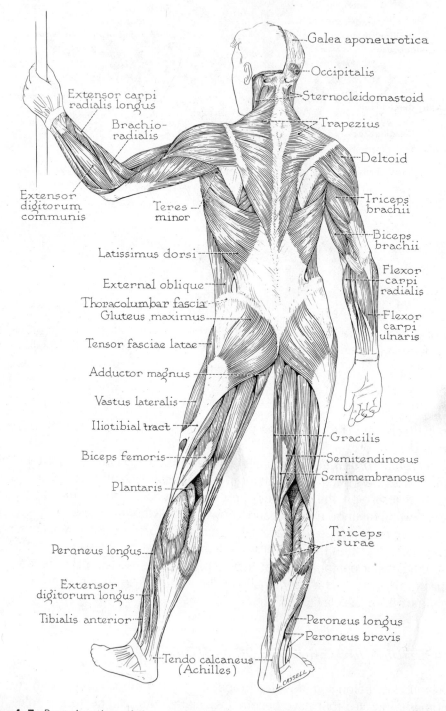

Figure 4–7 Posterior view of the muscles of a human being. (From King and Showers: Human Anatomy and Physiology. W.B. Saunders Company.)

on the clavicle and, together with the cleidobrachialis, protracts the arm. The sternocleidomastoid is also a single muscle in human beings (Fig. 4–5) but is subdivided in the cat. It acts primarily upon the head, turning it to the side and flexing it. If the head is fixed, parts of the complex may act on the clavicle. Clean off connective tissue from the ventral, lateral, and dorsal surfaces of the neck and from the dorsal part of the shoulder. You may have to remove more skin from the back of the head. Do not injure the large external jugular vein located superficially on the ventrolateral surface of the neck.

The cat has three divisions of the trapezius. The most caudal is the *thoracic trapezius,* a thin sheet of muscle covering the cranial part of the latissimus dorsi, from which it should be separated (Fig. 4–8). From their origin on the middorsal line of the thorax, the fibers of the thoracic trapezius converge to insert on the dorsal part of the scapular spine (Fig. 4–9).

A *cervical trapezius* lies cranial to the thoracic trapezius. It arises from the middorsal line of the front of the thorax and the back of the neck, and from an aponeurosis that interconnects the left and right cervical trapezius. Fibers of the cervical trapezius converge to insert on the ventral portion of the scapular spine and its metacromion process.

From their origin on the nuchal crest and middorsal line of the cranial part of the

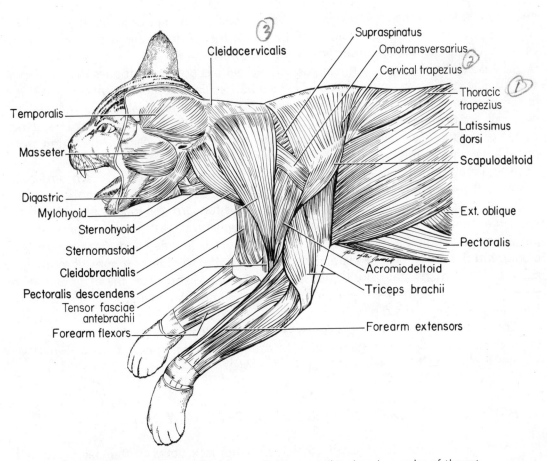

Figure 4–8 Lateral view of the superficial pectoral and neck muscles of the cat.

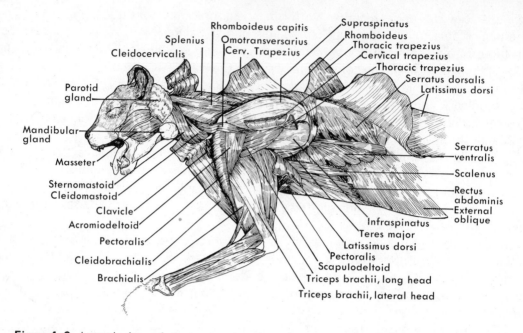

Figure 4–9 Lateral view of the pectoral and neck muscles of the cat after reflection of the trapezius group and latissimus dorsi.

neck, the fibers of the ***cleidocervicalis*** (clavotrapezius) extend caudally and ventrally to merge, at the level of the clavicle, with those of the cleidobrachialis, the muscle seen earlier covering the humeral insertion of the pectoralis.

The ***cleidobrachialis,*** which is a part of the deltoid group, continues distally from the clavicle to insert on the ulna. The cleidobrachialis, together with the cleidocervicalis, constitutes the ***brachiocephalicus.*** The cleidobrachialis and the cleidocervicalis are distinct muscles in vertebrates with a well-developed clavicle. They tend to merge as the clavicle becomes reduced in carnivores, but careful dissection may reveal a tendinous intersection between them.

There are two subdivisions of the sternocleidomastoid in the cat. A ***sternomastoid*** arises from the manubrium and extends cranially and dorsally to insert on the mastoid region of the skull (Figs. 4–6 and 4–8). As it extends forward, the muscle passes deep to the large external jugular vein. Muscle tissue superficial to the vein is part of the platysma and should be removed. The sternomastoid is a wide band that parallels the cranioventral border of the cleidocervicalis. Part of the insertion spreads onto the occipital region of the skull, and some authors distinguish these fibers as a sterno-occipitalis. Left and right sternomastoids of the cat merge near the sternum (Fig. 4–6) and should be cut apart.

The second subdivision, a ***cleidomastoid,*** extends from the clavicle (where it joins other muscles attaching on the clavicle) to the mastoid region of the skull (Figs. 4–6 and 4–9). Much of it lies deep to the sternomastoid.

(C) REMAINING SUPERFICIAL MUSCLES OF THE SHOULDER

The band of muscle that lies on the side of the neck between the cleidocervicalis and cervical trapezius is the ***omotransversarius.*** It arises from the metacromion process of the scapula ventral to the insertions of the cervical trapezius and extends forward to insert primarily on the transverse process of the atlas. Much of it lies deep to the cleidocervicalis (Fig. 4–9). The omotransversarius is a hypaxial muscle that pulls the scapula forward (protraction). We do not have this muscle.

The deltoid lies lateral to the shoulder joint, and most of it also lies ventral to the scapular spine. It is a single, deltoid-shaped muscle in human beings (Fig. 4–7), but is subdivided into three parts in the cat (Figs. 4–8 and 4–9). The *cleidobrachialis* (clavodeltoid) has been observed arising from the clavicle. It inserts on the ulna along with the brachialis (see the following section). The *acromiodeltoid* lies caudal to the cleidobrachialis. It arises from the acromion deep to the omotransversarius and inserts on the proximal portion of the humeral shaft. The *scapulodeltoid* lies along the lower border of the scapula. It arises from the scapular spine and inserts on the proximal end of the humerus. The cleidobrachialis protracts the arm, but the caudal parts of the complex are retractors and abductors of the humerus.

The *latissimus dorsi* has been observed on the side of the trunk caudal to the arm. In both cat and human beings it arises from the thoracolumbar fascia and from the spinous processes of the last thoracic vertebrae. From here it passes forward and ventrally to insert on the proximal end of the humerus in common with the teres major. In the cat, a small part of the muscle often forms a tendinous arc that encircles the muscles on the medial side of the arm and inserts with the pectoralis (Fig. 4–10). The latissimus dorsi retracts the humerus.

(D) DEEPER MUSCLES OF THE SHOULDER

Cut across the center of the latissimus dorsi at right angles to its fibers, and also across all the subdivisions of the trapezius muscle. Reflect the ends of these muscles, and clean out the fat and loose connective tissue from beneath them so as to expose the deeper muscles of the shoulder.

The supraspinous fossa of the scapula is occupied by the *supraspinatus* (Fig. 4–9). The muscle inserts on the greater tuberculum of the humerus and protracts the humerus. The infraspinous fossa is occupied by the *infraspinatus,* which inserts on the greater tuberculum of the humerus and rotates this bone outward. These muscles are the same in human beings.

The *teres major* arises from the caudal border of the scapula caudal and ventral to the infraspinatus. It passes forward to insert on the proximal end of the humerus in common with the primary insertion of the latissimus dorsi (Fig. 4–10). Its action is to rotate the humerus inward and to retract it. The human muscle is the same.

Dissect deeply between the cranial part of the infraspinatus and the long head of the triceps (the large muscle on the caudal surface of the brachium). You will eventually come upon a very small triangular muscle arising by a tendon from the cranial part of the caudal border of the scapula and inserting on the greater tuberculum of the humerus. This is the *teres minor;* it helps the infraspinatus rotate the humerus outward. This muscle is also present in human beings.

Examine the scapular region from a dorsal view. The large muscle that arises from the tops of the caudal cervical and cranial thoracic vertebrae and inserts along the dorsal margin of the scapula is the *rhomboideus* (Fig. 4–9). It is divided into a number of loosely associated bundles which give a coarse texture to the muscle. The most

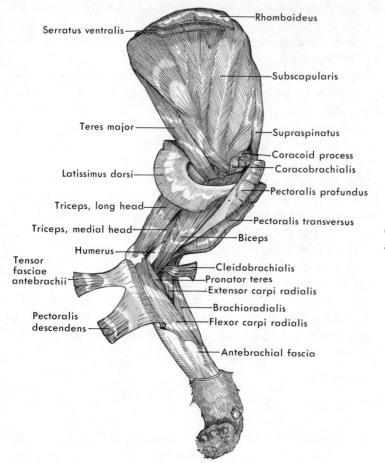

Serratus ventralis
Rhomboideus
Subscapularis
Teres major
Supraspinatus
Coracoid process
Coracobrachialis
Latissimus dorsi
Pectoralis profundus
Triceps, long head
Pectoralis transversus
Triceps, medial head
Biceps
Humerus
Tensor fasciae antebrachii
Cleidobrachialis
Pronator teres
Extensor carpi radialis
Brachioradialis
Pectoralis descendens
Flexor carpi radialis
Antebrachial fascia

Figure 4–10 Medial view of the muscles of the scapula and arm of the cat.

cranial bundle extends farther forward than the others to its origin from the back of the skull and is called the *rhomboideus capitis.* We lack this cranial slip, and the rest of the muscle is often subdivided into a more cranial rhomboideus minor and a caudal rhomboideus major. The rhomboideus is a hypaxial muscle that pulls the scapula toward the vertebrae, helps to hold it in place, and assists in its protraction and retraction.

Cut across the entire rhomboideus, pull the top of the scapula laterally, and clean away fat and loose connective tissue from the area exposed. The large fan-shaped muscle that you see is the *serratus ventralis* (Fig. 4–9). It arises by a number of slips from the ribs just dorsal to the junction of ribs and costal cartilages, and from the transverse processes of the posterior cervical vertebrae. It inserts on the dorsal border of the scapula ventral to the insertion of the rhomboideus. The serratus ventralis is a hypaxial muscle. In quadrupeds it forms, together with the pectoralis, a muscular sling that transfers much of the weight of the body to the pectoral girdle and appendage. The serratus ventralis is the major component in the sling (Fig. 4–11).

We have a similar muscle called the serratus anterior (Fig. 4–5) that arises from the ribs and extends to the dorsal border of the scapula. It pulls the scapula toward the front of the chest, an action that occurs in pushing. The cranial part of the cat muscle,

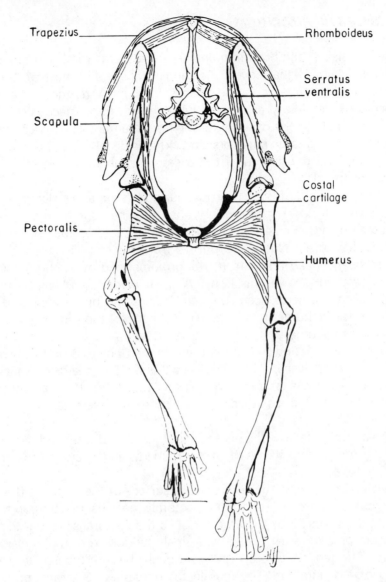

Figure 4–11 Diagrammatic anterior view of a section through the thorax at the level of the pectoral girdle to show muscular connections between the trunk and appendage.

which arises from the cervical vertebrae, is represented in human beings by a distinct muscle known as the levator scapulae whose action is to pull the scapula toward the head.

Clean out the area between the serratus ventralis and the medial surface of the scapula. The subscapular fossa is occupied by a large, pinnated *subscapularis* (Fig. 4–10), which passes to insert on the lesser tuberculum of the humerus. This muscle pulls the humerus medially (adduction). The human muscle is the same.

(E) MUSCLES OF THE BRACHIUM

Clean the muscles of the brachium and separate them from each other as much as possible. The large muscle that covers the caudal surface and most of the medial and lateral surface of the humerus is the triceps brachii. A *tensor fasciae antebrachii* (dorsoepitrochlearis) is closely associated with the medial surface of the triceps and should be studied first. In the cat, the tensor fasciae antebrachii arises primarily from the lateral surface of the latissimus dorsi and extends distally along the median surface of the arm to insert on the tendon of the triceps and on the antebrachial fascia (Fig. 4–6). We lack this muscle.

Cut through the origin of the tensor fasciae antebrachii and reflect the muscle. The *triceps brachii,* which is now exposed clearly, has three heads in both cat (Figs. 4–9 and 4–10) and human beings (Fig. 4–7). The *long head,* located on the posterior surface of the humerus, is the largest. Note that it arises from the scapula caudal to the glenoid cavity. A large *lateral head* arises from the proximal end of the humerus and covers much of the lateral surface of this bone. A small *medial head* can be found on the medial surface of the humerus deep to several nerves and blood vessels. It arises from most of the shaft of the humerus (Fig. 4–10). All parts of the triceps insert in common on the olecranon of the ulna.

A small *anconeus* is found in both cat and human beings deep to the distal end of the lateral head of the triceps. Cut and reflect this head to see it. The anconeus arises from the lateral epicondyle of the humerus and inserts on the olecranon beside the insertion of the triceps. Triceps, anconeus, and tensor fasciae antebrachii are extensors of the forearm.

The anterolateral surface of the humerus is covered in the cat and human beings by the *brachialis.* It arises from the shaft of the humerus and inserts on the proximal end of the ulna (Fig. 4–9).

The *biceps brachii* lies on the anteromedial surface of the humerus (Fig. 4–10). To see it clearly, cut through and reflect the pectoralis near its insertion on the humerus. (The mass of nerves passing to the arm dorsal to the pectoralis belongs to the brachial plexus.) The biceps of the cat arises by a single tendon from the dorsal edge of the glenoid cavity. The tendon passes distally through the intertubercular groove on the proximal end of the humerus. The muscle then enlarges to form a prominent belly adjacent to the ventral surface of the humeral shaft, and finally inserts by a short tendon on the radial tuberosity. The human biceps (Fig. 4–5) also has a second, shorter head that arises from the coracoid process. Biceps and brachialis are the major flexors of the forearm. The biceps also assists in supination of the forearm.

Clean off connective tissue from the medial side of the shoulder joint; you will find a short band of muscle arising from the coracoid process and inserting on the proximal end of the humerus. This is the *coracobrachialis* (Fig. 4–10). It helps pull the humerus toward the body (adduction). In some individuals, a long portion of the coracobrachialis passes to the distal end of the humeral shaft. We have a similar muscle.

(F) MUSCLES OF THE FOREARM

Before the muscles of the forearm and hand can be studied it is necessary to remove the very extensive *antebrachial fascia.* The deeper part of this fascia is

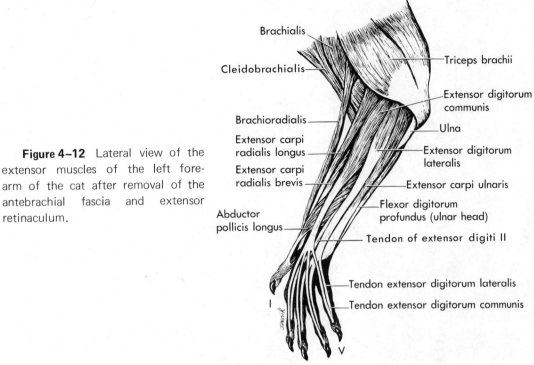

Figure 4–12 Lateral view of the extensor muscles of the left forearm of the cat after removal of the antebrachial fascia and extensor retinaculum.

Labels in figure:
Brachialis
Cleidobrachialis
Brachioradialis
Extensor carpi radialis longus
Extensor carpi radialis brevis
Abductor pollicis longus
Triceps brachii
Extensor digitorum communis
Ulna
Extensor digitorum lateralis
Extensor carpi ulnaris
Flexor digitorum profundus (ulnar head)
Tendon of extensor digiti II
Tendon extensor digitorum lateralis
Tendon extensor digitorum communis

continuous with the tendons of the pectoralis descendens, the tensor fasciae antebrachii, and the triceps. It forms a dense fibrous sheet which dips down between many of the muscles and also attaches onto the ulna and radius. At the level of the wrist, part of this fascia forms ligaments which encircle the wrist and hold the muscle tendons in place. A band of dense fibers, the *extensor retinaculum,* bridges the tendon grooves on the dorsal or extensor surface of the radius, and a comparable *flexor retinaculum* crosses the palmar side of the wrist. This antebrachial fascia continues into the hand and, on the palmar side, helps to form fibrous sheaths, the *vaginal ligaments,* through which the flexor tendons of the fingers run. After the fascia has been removed, separate the major muscles before attempting to identify them.

The muscles of the forearm and hand can be sorted into an extensor group located on the anterolateral surface of the forearm and the back of the hand, when the hand is in the prone position, and a flexor group located on the posteromedial surface of the forearm and palm of the hand. In the elbow region, the insertions of the biceps, brachialis, and cleidobrachialis pass between these groups at the anterior border of the arm, and the ulna and insertion of the triceps separate them posteriorly.

Most of the extensors arise from or near the lateral epicondyle of the humerus. In the cat one can recognize the following superficial muscles at the level of the elbow and beginning from the medial or thumb side of the forearm (Fig. 4–12): a brachioradialis, an extensor carpi radialis complex, an extensor digitorum communis, an extensor digitorum lateralis, and an extensor carpi ulnaris. Most of these muscles are the same in human beings (Figs. 4-13 and 4–14). The *brachioradialis* arises more proximally from the humerus than the others and inserts on the styloid process of the radius. The extensor carpi radialis complex can be divided into a more superficial and anterior *extensor carpi radialis longus* and a deeper and more posterior *extensor carpi*

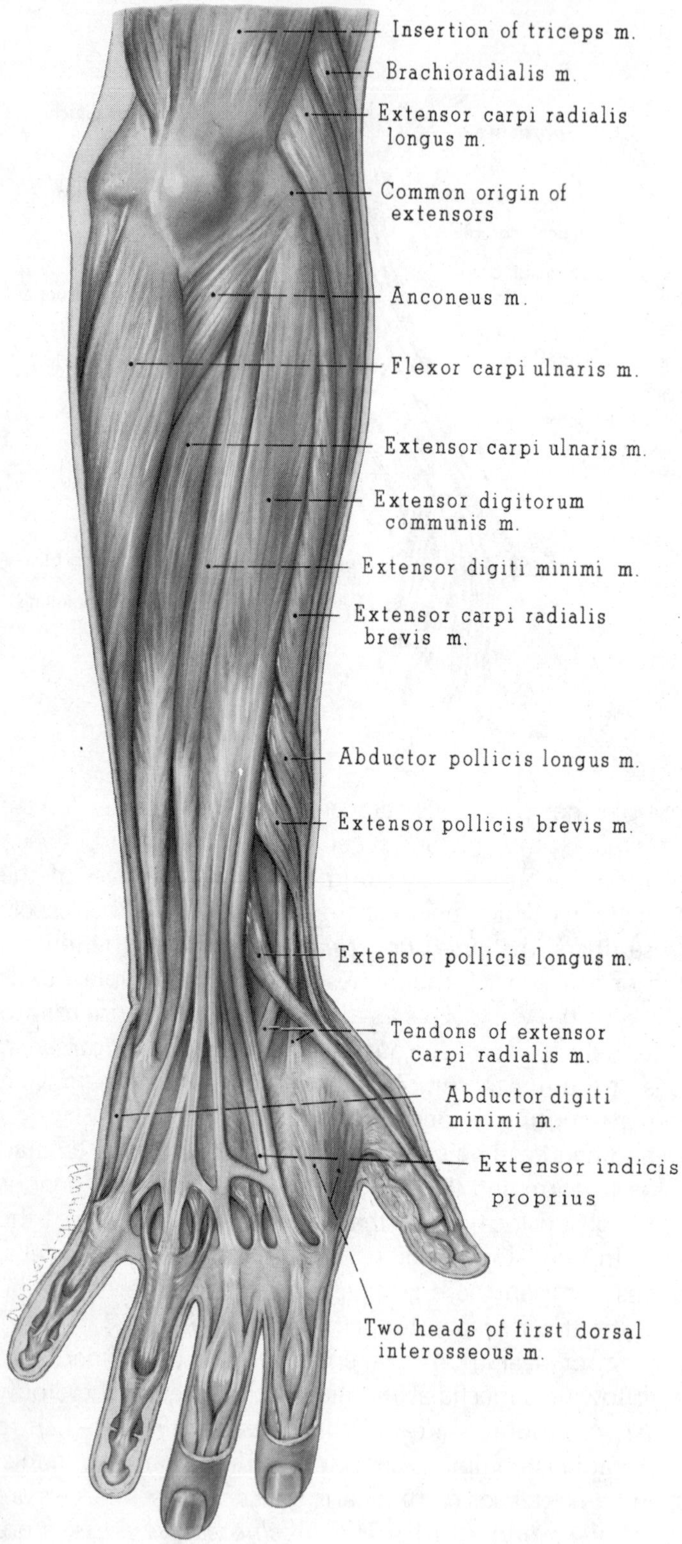

Insertion of triceps m.

Brachioradialis m.

Extensor carpi radialis longus m.

Common origin of extensors

Anconeus m.

Flexor carpi ulnaris m.

Extensor carpi ulnaris m.

Extensor digitorum communis m.

Extensor digiti minimi m.

Extensor carpi radialis brevis m.

Abductor pollicis longus m.

Extensor pollicis brevis m.

Extensor pollicis longus m.

Tendons of extensor carpi radialis m.

Abductor digiti minimi m.

Extensor indicis proprius

Two heads of first dorsal interosseous m.

Figure 4-13 Muscles of the extensor surface of the human right forearm and hand. (From Jacob and Francone: Structure and Function in Man. W.B. Saunders Company.)

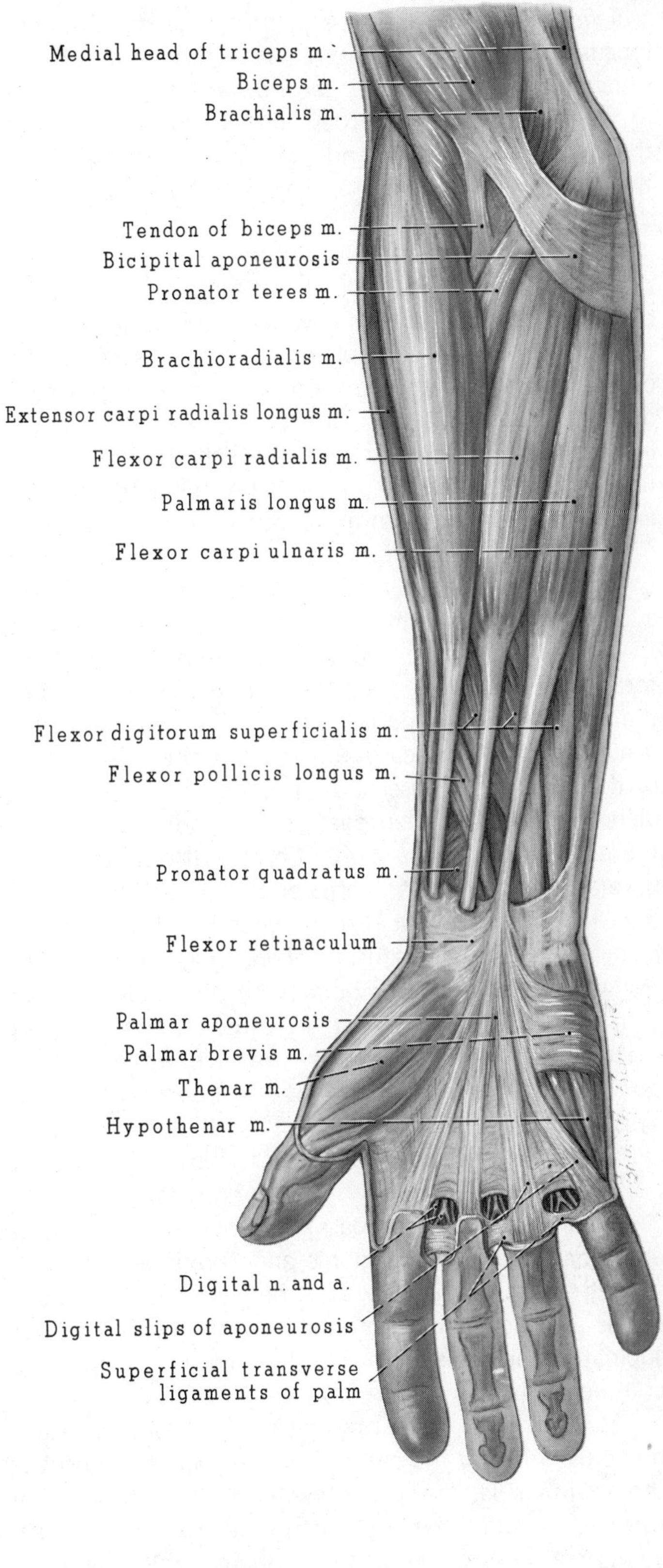

Medial head of triceps m.
Biceps m.
Brachialis m.

Tendon of biceps m.
Bicipital aponeurosis
Pronator teres m.

Brachioradialis m.

Extensor carpi radialis longus m.

Flexor carpi radialis m.

Palmaris longus m.

Flexor carpi ulnaris m.

Figure 4–14 Muscles of the flexor surface of the human right forearm and hand. (From Jacob and Francone: Structure and Function in Man. W.B. Saunders Company.)

Flexor digitorum superficialis m.
Flexor pollicis longus m.

Pronator quadratus m.

Flexor retinaculum

Palmar aponeurosis
Palmar brevis m.
Thenar m.
Hypothenar m.

Digital n. and a.
Digital slips of aponeurosis
Superficial transverse ligaments of palm

radialis brevis. The tendons of both pass deep to the tendon of the abductor pollicis longus (see below) to insert upon the bases of the second and third metacarpals, respectively. The *extensor carpi ulnaris* has a comparable position on the ulnar border of the forearm, and its broad tendon inserts upon the base of the fifth metatarsal. All these muscles act to extend the forearm and hand, and the brachioradialis also assists the supinator (see below) in rotating the radius in such a way that the palm is turned dorsally (supination).

The long digital extensors are more complex. After passing through a groove on the dorsal surface of the radius, the tendon of the cat's *extensor digitorum communis* breaks up into four tendons which pass along the dorsal surface of digits 2 to 5. The tendons are bound to each phalanx of the digits by connective tissue, but the most conspicuous attachment is on the terminal phalanges. The tendon of the *extensor digitorum lateralis* also passes through a groove on the radius and divides at about the level of the wrist into three or four parts which pass down the dorsal surface of digits 2 to 5 or 3 to 5. These tendons at first lie somewhat on the radial side of those of the extensor digitorum communis but eventually unite with those of the latter muscle. These two muscles are extensors of digits 2 to 5. The human extensor digitorum communis is essentially the same, but we lack the extensor digitorum lateralis and have a long extensor digiti minimi inserting on the little finger (Fig. 4–13).

Three muscles lie deep to the preceding muscles on the extensor surface of the forearm. They can be seen by cutting and reflecting the two long digital extensors and the extensor carpi ulnaris. A narrow *extensor digiti II* (extensor indicis proprius) arises deep to the extensor carpi ulnaris from the proximal three fourths of the lateral surface of the ulna, and it inserts by a tendon that goes to the middle phalanx of the second digit (Fig. 4–12). Often part of the tendon passes to the thumb. The muscle assists in the extension of these digits. Certain of our finger muscles are more complex, for we have a grasping hand and greater control over finger movements than a cat. The part of the cat's extensor digiti II that sometimes inserts on the thumb is represented in human beings by a distinct extensor pollicis longus (Fig. 4–13).

A powerful *abductor pollicis longus* arises from much of the lateral surface of the ulna and adjacent parts of the radius. Its fibers converge, go deep to the tendon of the extensor digitorum communis, and form a tendon that goes superficial to the extensor carpi radialis tendons (Fig. 4–12) to insert upon the radial side of the first metacarpal. This muscle extends and abducts the thumb. The single abductor pollicis longus of the cat is subdivided in human beings into an abductor pollicis longus and an extensor pollicis group.

Finally, a *supinator* passes obliquely across the radius deep to the belly of the extensor digitorum communis and proximal to the abductor pollicis longus. It arises from the lateral epicondyle of the humerus and elbow ligaments and inserts upon the radius. Its diagonal fibers enable it to act as a powerful supinator of the hand. Our supinator originates from the ulna. The small muscles confined to the hand will not be considered.

Most of the flexor muscles arise from or near the medial epicondyle of the humerus. In both the cat and human beings, the superficial muscles are, beginning at the thumb side, the pronator teres, the flexor carpi radialis, the flexor digitorum superficialis, and the flexor carpi ulnaris (Figs. 4–14 and 4–15). We often have an additional muscle, the palmaris longus, superficial to the flexor digitorum superficialis. The *pronator teres* passes diagonally from the medial epicondyle to the medial border

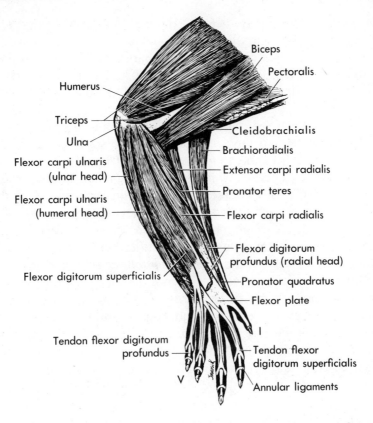

Figure 4–15 Medial view of the flexor muscles of the forearm of the cat after removal of the antebrachial fascia, flexor retinaculum, and palmar aponeurosis.

of the radius, and its action (pronation) rotates the forearm in such a way that the palm faces the ground. Part of the origin of this muscle in human beings is on the ulna.

The *flexor carpi radialis* extends down the radial border of the forearm and forms a long tendon which passes deep to other tendons in the hand to insert on the proximal ends of the second and third metacarpals. The *flexor carpi ulnaris* has a comparable position on the ulnar border of the forearm. It arises by two heads, one from the medial epicondyle and one from the surface of the olecranon of the ulna; it inserts upon the pisiform bone. The chief action of these two muscles is to flex the hand at the wrist.

The *flexor digitorum superficialis,* widest of the superficial muscles, lies between the two carpal flexors. It arises partly from the medial epicondyle of the humerus and partly from the surface of a deeper muscle (flexor digitorum profundus). At the level of the wrist it forms a tendon that passes deep to the flexor retinaculum. Part of this tendon has cutaneous attachments to the foot pads, but most divides into four tendons that go to the bases of the second through fifth fingers. Here each tendon splits and attaches onto the side of the middle phalanx of the digit. The tendons of the flexor digitorum profundus pass through the openings formed by the splitting of the superficialis tendons. This muscle flexes the digits near their middle. It originates from the humerus and radius in human beings.

Cut and reflect the flexor carpi radialis, flexor digitorum superficialis, and flexor carpi ulnaris. The muscle complex deep to them is the *flexor digitorum profundus.* It consists of three heads of origin all of which converge at the wrist to form a powerful flexor plate. The first, or ulnar head, arises from most of the length of the outer border of the ulna, and part of it is visible from the extensor side of the forearm (Fig. 4–12); the second, or humeral head, arises from the medial epicondyle deep to the origin of the palmaris longus and extensor carpi ulnaris; the third, or radial head (Fig. 4–15), is the deepest and arises from the middle third of the radius, the interosseous membrane stretching between radius and ulna, and from the adjacent parts of the ulna. Their common tendon extends into the palm as a powerful flexor plate and then breaks up into five strong flexor tendons which run through ligamentous sheaths, beneath annular ligaments, and finally insert on the terminal phalanges of the digits. The tendons on the second-to-last finger perforate the tendons of the flexor digitorum superficialis. This muscle flexes all segments of the digits. The part of the muscle going to thumb has separated in human beings as a distinct flexor pollicis longus (Fig. 4–14). Certain small intrinsic muscles in the hand (the *lumbricales*) arise superficially from the flexor plate. Other intrinsic hand muscles are not described.

Separate the tendon of the flexor carpi radialis, which goes deep to the flexor plate, and the radial head of the flexor digitorum profundus. The very deeply situated muscle which you see is the *pronator quadratus* (Fig. 4–15). Its fibers pass diagonally from their origin on the distal third of the ulna to their insertion on the outer border of the radius. It assists in hand pronation.

PELVIC MUSCLES

Clean off the fat and superficial fascia from the surfaces of the pelvic region and thigh as a preliminary to the dissection. Be sure to remove the large wads of fat lateral to the base of the tail and from the depression *(popliteal fossa)* posterior to the knee. Start to separate the more obvious muscles.

(A) LATERAL THIGH AND ADJACENT MUSCLES

The most cranial muscle on the thigh of the cat is the *sartorius* (Fig. 4–16). It is a band extending from the crest and ventral border of the ilium (its origin) to the patella and medial side of the tibia (its insertion). Most of the muscle lies on the medial surface of the thigh, but part of it can be seen laterally. The muscle adducts and rotates the femur and extends the shank. The human sartorius is very similar (Figs. 4–5 and 4–7).

Caudal to it on the lateral surface of the thigh of both cat and human beings there is a tough, white fascia—the *fascia lata.* A triangular muscle mass *(tensor fasciae latae)* arises from the ventral border of the ilium and from the surface of adjacent muscles. It passes to insert in the dorsal part of the fascia lata, and it acts to tighten this fascia and to extend the shank.

The very broad muscle caudal to the fascia lata in the cat is the *biceps femoris.* It arises by a single head from the tuberosity of the ischium and then fans out to insert by a broad aponeurosis that extends from the patella to the middle of the shaft of the tibia. The muscle forms the lateral wall of the popliteal fossa. Its action is to flex the

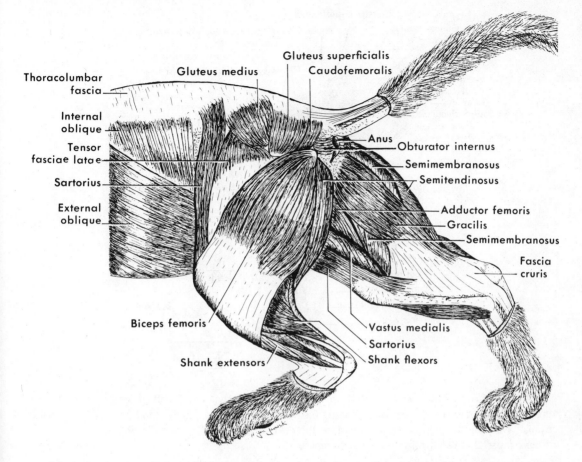

Figure 4–16 Superficial pelvic and thigh muscles of the cat. Lateral muscles can be seen on the left leg and medial muscles on the right leg.

shank and abduct the thigh. Our biceps femoris has shifted more onto the posterior surface of the thigh (Fig. 4–7), so it also acts to retract the thigh. It has a second head of origin from the femur and is not so broad a muscle; its insertion is limited to the proximal end of the tibia. You can feel its powerful tendon on the lateral side of the popliteal fossa.

The stout muscle arising from the ischial tuberosity, caudal to the origin of the biceps femoris, is the *semitendinosus.* It inserts on the medial side of the distal end of the tibia (Fig. 4–18). It acts to flex the shank and retract the thigh. It has a similar position in human beings, and its powerful tendon of insertion can be felt on the posteromedial border of the popliteal fossa.

The band of muscle cranial and dorsal to the origin of the biceps femoris is the *caudofemoralis* (coccygeofemoralis). It arises from the second and third caudal vertebrae, passes beneath the anterior border of the biceps femoris, and forms a thin tendon that inserts on the patella. Lift up the biceps to follow it. The caudofemoralis abducts the thigh and extends the shank. This muscle appears to have evolved from a part of the gluteus superficialis and, for this reason, is sometimes called the gluteobiceps. It is absent in human beings.

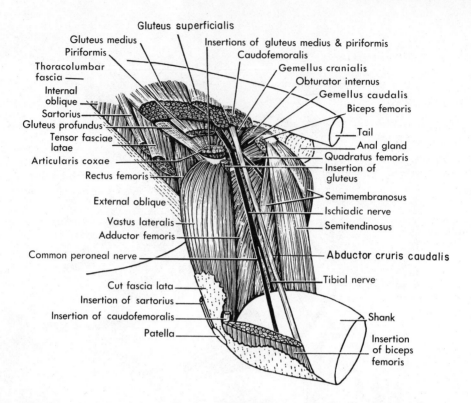

Figure 4–17 Lateral view of the deep muscles of the pelvis and thigh of a cat. The sartorius, tensor fasciae latae, biceps femoris, gluteus medius, gluteus superficialis, caudofemoralis, and piriformis have been largely cut away to expose the deeper muscles. However, the origins and insertions of these muscles are shown as points of reference.

Cut across the biceps and the caudofemoralis near their origins, being careful not to cut a very slender band of muscle that lies beneath them. This band, which will be seen on reflecting the biceps, is the *abductor cruris caudalis* (Fig. 4–17). It arises from the second caudal vertebra, inserts on the tibia with the biceps femoris, and assists this muscle in abduction of the thigh and flexion of the shank. The abductor cruris caudalis is closely associated with the biceps femoris and has probably evolved from it. We do not have this muscle.

(B) GLUTEAL COMPLEX AND DEEPER PELVIC MUSCLES

The muscle mass covering the dorsolateral surface of the sacrum between the caudofemoralis and the sartorius is the gluteus complex (Fig. 4–16). Carefully remove overlying connective tissue and fascia, and start to separate the layers of the complex. The most superficial part is the gluteus superficialis; the next deeper layer, the gluteus medius. In the cat the superficialis also lies somewhat caudal and distal to the medius. The gluteus superficialis is homologous to the human gluteus maximus, but it is usually not so large as the medius in quadrupeds.

The *gluteus superficialis* arises from the sacral fascia and from the spinous processes

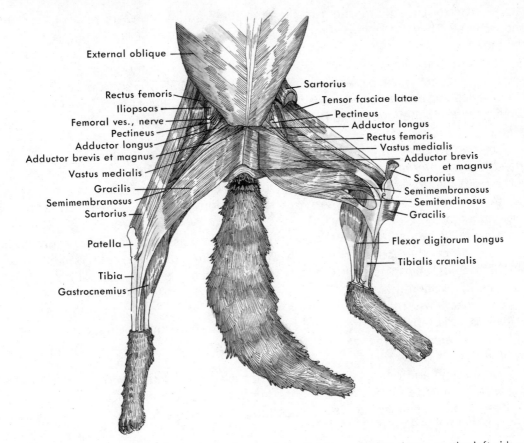

Figure 4–18 Medial thigh muscles of the cat. Superficial muscles are shown on the left side of the drawing; deeper ones, after reflection of the sartorius and gracilis, on the right side.

of sacral and anterior caudal vertebrae. Fibers converge to insert on the greater trochanter of the femur.

Cut through the belly of the gluteus superficialis and reflect its ends. The *gluteus medius,* which lies partly deep to the superficialis, arises from the crest and lateral surface of the ilium and adjacent vertebrae. It inserts upon the greater trochanter. Both gluteus superficialis and gluteus medius act primarily as thigh abductors in quadrupeds. The human gluteus maximus is more posteriorly situated and plays a major role in holding the trunk erect over the pelvis and the hind legs.

A thin *piriformis* lies deep to the caudal portion of the gluteus medius, from which it cannot always be separated easily (Fig. 4–17). It arises from the last sacral and first caudal vertebrae and inserts with the medius on the greater trochanter. It too abducts the thigh.

Cut through and reflect the piriformis if this was not done with the reflection of the gluteus medius. The prominent *ischiadic nerve* lies deep to the piriformis. A cranial *gluteus profundus* (gluteus minimus of human anatomy) and a more caudal *gemellus cranialis* (superior) lie deep to the gluteus medius and piriformis. The ischiadic nerve crosses the latter muscle. These two muscles are more or less united, and are somewhat difficult to separate except at their origins. The gluteus profundus arises from the lateral surface of the ilium; the gemellus cranialis, from the dorsal borders of the ilium and ischium. They insert in common on the greater trochanter and rotate and abduct the thigh.

Look deep to the cranioventral border of the gluteus profundus and between the origins of the rectus femoris and vastus lateralis (Fig. 4–17). The short muscle observed is the *articularis coxae.* It arises from part of the lateral iliac surface, inserts upon the proximal end of the femur, and helps to flex the thigh. This muscle is absent in human beings.

The narrow band of muscle located just caudal to the gemellus cranialis is a part of the *obturator internus* (Fig. 4–17). The obturator internus arises on the inside of the pelvis from the borders of the obturator foramen and passes over the dorsal rim of the ischium (where it can now be seen) to its insertion in the trochanteric fossa. It is a thigh abductor.

A *gemellus caudalis* (inferior) lies caudal and partly deep to the obturator internus. It arises from the lateral surface of the ischium cranial to the origin of the caudofemoralis, and inserts with the obturator internus in the trochanteric fossa. It is a thigh abductor and retractor.

The *quadratus femoris* is the rather thick band of muscle distal to the gemellus caudalis. It arises from the ischial tuberosity deep to the origin of the biceps femoris, and passes to insert on the femur at the bases of the greater and lesser trochanters. It is primarily a thigh retractor.

Cut through and reflect the gemellus caudalis and quadratus femoris. An *obturator externus* lies deep to them. It arises on the lateral surface of the pelvis from the borders of the obturator foramen and inserts deep in the trochanteric fossa. Its action is thigh rotation and retraction.

(C) QUADRICEPS FEMORIS COMPLEX

The front portion of the thigh of mammals is covered by a group of four muscles which insert in common on the patella and *patella tendon,* which, in turn, attaches to the tuberosity of the tibia. The whole complex is often referred to as the *quadriceps femoris,* and it is the primary shank extensor. The patella permits the common tendon of these muscles to slide easily across the knee joint.

Cut across the belly of the sartorius and across the fascia lata at right angles to its fibers. Make any further cuts necessary to reflect the fascia lata and its tensor fasciae latae. Clean the area exposed. The large muscle on the craniolateral surface of the thigh, which was covered by the fascia lata, is the *vastus lateralis* (Fig. 4–17). It arises from the greater trochanter and shaft of the femur. The narrower muscle on the front of the thigh, medial to the vastus lateralis, is the *rectus femoris* (Fig. 4–18). Since it arises from the ilium just cranial to the acetabulum, it also acts across the hip and helps to protract the thigh. The muscle on the medial surface of the thigh, caudal to the rectus femoris, is the *vastus medialis* (Fig. 4–18). It arises from the shaft of the femur. Both the rectus femoris and the vastus medialis lie beneath the sartorius. Pull the rectus femoris and vastus lateralis apart. The deep muscle observed between them is the *vastus intermedius.* It, too, arises from the shaft of the femur. This group of muscles is the same in human beings (Fig. 4–5).

(D) CAUDOMEDIAL THIGH MUSCLES

The medial surface of the thigh caudal to the sartorius and quadriceps femoris is largely covered by the *gracilis* (Fig. 4–18), a broad, thin muscle that arises from the pubic and ischial symphyses. It inserts by an aponeurosis onto the tibia and crural fascia. The gracilis adducts and retracts the thigh and flexes the crus.

Cut through the gracilis and reflect its ends. Notice again the semitendinosus seen earlier from the lateral surface. It inserts on the tibia and forms much of the medial wall of the popliteal fossa. The broad, thick muscle lying deep to the gracilis and cranial to the semitendinosus is the **semimembranosus.** It arises from the tuberosity and caudal border of the ischium, and inserts upon the medial epicondyle of the femur and the adjacent part of the tibia. It is a retractor of the thigh and flexor of the shank.

A small, triangular-shaped **pectineus** arises from the cranial border of the pubis just caudal to the point where the femoral blood vessels emerge from the body wall. It inserts on the femoral body beside the origin of the vastus medialis. It is a thigh adductor.

An adductor group of muscles lies between the pectineus and semimembranosus, and these muscles take their origin from the cranial border of the pubis and from the pubic and ischial symphyses. They insert along the femoral body and adduct the thigh. The group is subdivided somewhat differently in different mammals. In the cat, the slender, cranial component is an *adductor longus,* and the rest constitutes an *adductor brevis et magnus.*

(E) ILIOPSOAS COMPLEX AND ADJACENT MUSCLES

Note the thick bundle of muscle that emerges from the body wall medial to the origin of the rectus femoris. This is the iliopsoas complex. It may be studied now, or its study may be postponed until the abdominal viscera have been dissected. If the former choice is made, trace the bundle forward by cutting through the muscle layers of the abdominal wall. The complex lies in a retroperitoneal position. The main part of the bundle represents the *psoas major.* This portion arises primarily from the bodies of the last two or three thoracic vertebrae and from the bodies of all the lumbar vertebrae. It inserts on the lesser trochanter and protracts and rotates the thigh. Lateral and slightly dorsal to its extreme caudal portion you will see a group of fibers arising from the ventral border of the ilium. These fibers represent the *iliacus* of human beings and certain other mammals. In the cat, psoas major and iliacus are more or less united with each other. They insert and act in common and may be called the *iliopsoas* (Fig. 8–12, p. 160).

The thin muscle medial to the psoas major is the *psoas minor.* It arises from the bodies of the caudal thoracic and cranial lumbar vertebrae. It inserts on the girdle near the origin of the pectineus. The psoas minor is one of the subvertebral hypaxial muscles referred to on page 47, rather than an appendicular muscle. Its action is flexion of the back. This small muscle is often absent in human beings.

Lift up the lateral border of the psoas major near its middle. The thin muscle that lies on the ventral surface of the transverse processes is the *quadratus lumborum,* another subvertebral hypaxial muscle. It arises primarily from the bodies and transverse processes of the lumbar and last several thoracic vertebrae, but some of its fibers also spring from the last rib. It inserts on the ilium cranial to the origin of the iliacus. Its action is lateral flexion of the vertebral column. We have a similar muscle.

(F) MUSCLES OF THE SHANK

The shank is covered by a tough *fascia cruris,* which is partly united with the tendons of certain thigh muscles, including the biceps femoris and gracilis. Remove the fascia and reflect thigh muscles inserting on the shank, but try to leave their tendons intact. Separate the more obvious shank muscles.

As was the case in the forearm and hand, the muscles of the shank and foot fall into extensor and flexor groups, but the groups are not quite so clearly separated as in the forearm. The extensors lie on the craniolateral surface of the shank; the flexors, on the caudomedial surface. They are separated by an exposed strip of the tibia on the medial side (Figs. 4–21 and 4–22), and the position of the fibula indicates their separation laterally, although the fibula is not exposed at the surface (Figs. 4–19 and 4–20).

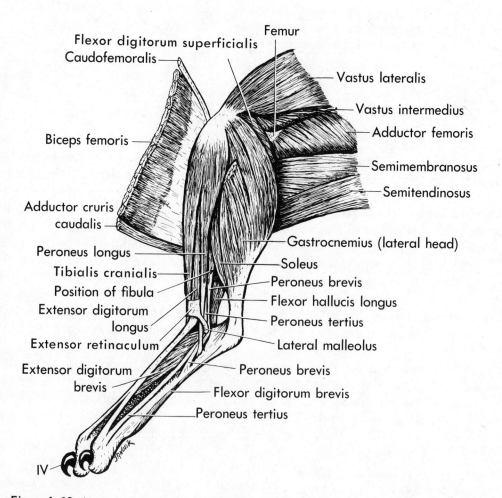

Figure 4–19 Lateral view of the left shank muscles of the cat after reflection of the biceps femoris and removal of the crural fascia.

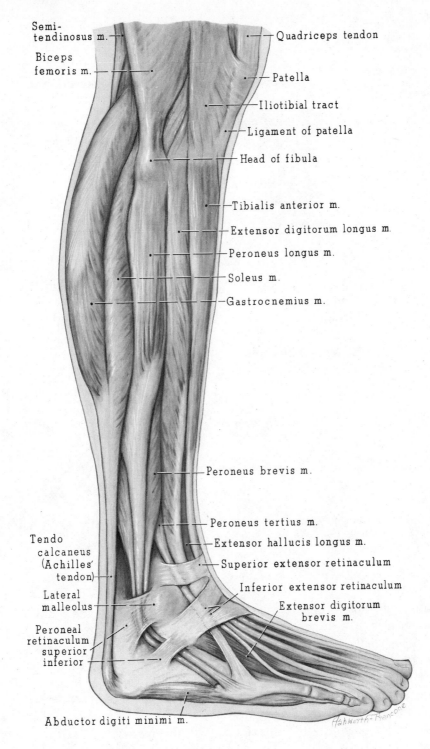

Semi-
tendinosus m.
Biceps
femoris m.

Quadriceps tendon
Patella
Iliotibial tract
Ligament of patella
Head of fibula
Tibialis anterior m.
Extensor digitorum longus m.
Peroneus longus m.
Soleus m.
Gastrocnemius m.

Peroneus brevis m.

Peroneus tertius m.
Extensor hallucis longus m.
Superior extensor retinaculum
Inferior extensor retinaculum
Extensor digitorum
brevis m.

Tendo
calcaneus
(Achilles'
tendon)
Lateral
malleolus
Peroneal
retinaculum
superior
inferior

Abductor digiti minimi m.

Figure 4–20 Lateral view of the right shank and foot muscles of a human being. (From Jacob and Francone: Structure and Function in Man. W.B. Saunders Company.)

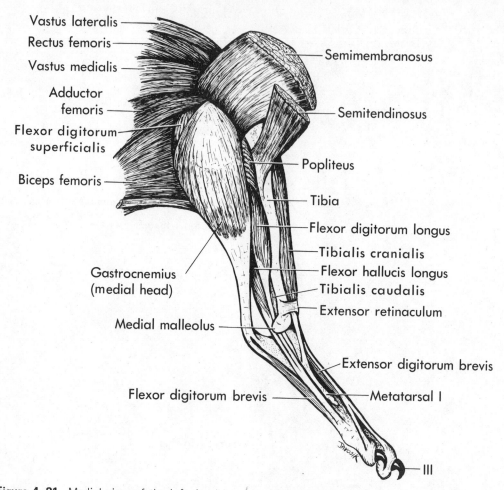

Figure 4–21 Medial view of the left shank muscles of the cat after reflection of the tensor fascia lata, gracilis, semitendinosus and semimembranosus, and removal of the crural fascia.

The large calf muscle on the caudal or posterior surface of the shank in both cat and human beings is a functional unit composed of the gastrocnemius, the flexor digitorum superficialis, and the soleus (Figs. 4–19 to 4–22). The lateral head of the *gastrocnemius* arises from the lateral epicondyle of the femur, lateral surface of the patella and adjacent parts of the tibia, and by a small slip from the crural fascia. Its medial head arises from the medial epicondyle of the femur. A *flexor digitorum superficialis* (plantaris) is situated between the two heads of the gastrocnemius, where it takes its origin deep to the lateral head of the gastrocnemius from the lateral epicondyle of the femur and adjacent part of the patella. The fleshy part of the *soleus* lies deep to the distal portion of the lateral head of the gastrocnemius. It arises from the proximal one third of the fibula. All converge to form a large common tendon, the *calcaneus tendon* (Achilles tendon), which inserts upon the calcaneus. From an evolutionary viewpoint these muscles belong to the ventral limb musculature of flexors, and their action upon the foot is one of plantar flexion. This action is often

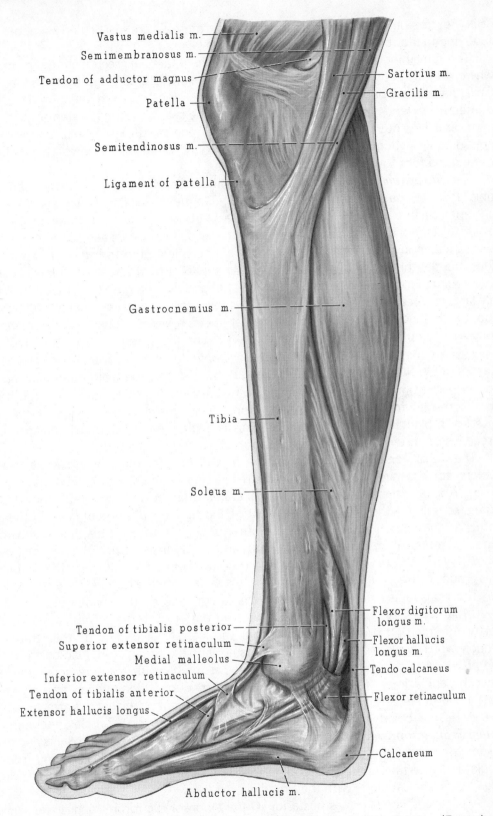

Figure 4–22 Medial view of the right shank and foot muscles of a human being. (From Jacob and Francone: Structure and Function in Man. W.B. Saunders Company.)

called extension in human anatomy. These muscles are particularly important in thrusting the foot upon the ground, hence their large size.

The remaining four flexor muscles of the shank are best seen from the medial side where they lie between the tibia and the group just described (Figs. 4–21 and 4–22). A triangular-shaped *popliteus* arises by a narrow tendon from the lateral epicondyle of the femur, extends toward the medial side of the shank, passing posterior to the knee joint, and fans out to insert upon the proximal one third of the tibia. Its insertion is not so long in human beings. It helps to flex the shank and also rotates it slightly, turning the foot toward the midventral line.

The *flexor digitorum longus* arises from the head of the fibula and the shaft of the tibia distal to the insertion of the popliteus. The *flexor hallucis longus* arises from the posterior surface of much of the rest of the fibula and tibia. Both these muscles form powerful tendons which pass posterior to the medial malleolus and soon unite to form a broad tendon plate which covers much of the sole of the foot deep to an intrinsic foot muscle, the *flexor digitorum brevis.* At the level of the toes this plate breaks up into four tendons which extend down the flexor side of digits 2 to 5 to the terminal phalanges. These two muscles flex the toes and assist in plantar flexion of the whole foot. In human beings, in which the great toe is well developed, the flexor hallucis longus is completely independent and inserts upon the distal phalanx of the first digit.

Last of the flexors is a small *tibialis caudalis* (posterior), which lies between the flexors digitorum longus and hallucis longus. It arises in part from the fibula and tibia and in part from the surface of the adjacent muscles. Its tendon, too, passes posterior to the medial malleolus and inserts upon certain distal tarsals. It assists the other muscles in plantar flexion of the foot. This is a larger muscle in human beings, and its broad insertion tendon also helps to support the arches of the foot.

The dorsal, or extensor, musculature of the shank is much less massive than the flexors, for it is involved primarily in the recovery movements of the limb. Most cranial or anterior of the extensors is the *tibialis cranialis* (anterior) (Figs. 4–20 and 4–21). It arises from about the proximal one third of the fibula and adjacent parts of the tibia, forms a long tendon which crosses the anterior surface of the tibia, goes beneath the extensor retinaculum near the ankle joint, and finally inserts upon the first metatarsal, which is large in our case but reduced in the cat. Together with many other muscles in this group it extends the foot—a motion often called dorsiflexion.

The remaining extensors all lie on the lateral side of the shank (Figs. 4–19 and 4–20). An *extensor digitorum longus* is located posterior to the tibialis cranialis, which partly covers it. It arises in the cat from the lateral epicondyle of the femur by a tendon which traverses the knee joint capsule, but the origin in human beings is from the proximal two-thirds of the fibula. After passing beneath the extensor retinaculum near the ankle joint, the muscle breaks up into four tendons which pass down the dorsum of digits 2 to 5, finally inserting on the terminal phalanges. Its attachments on the digits are closely associated with those of intrinsic foot muscles, including the *extensor digitorum brevis.* The extensor digitorum longus extends the digits and assists in dorsiflexion of the foot. We also have an extensor hallucis longus arising from the fibula and inserting upon the terminal phalanx of the great toe, but this muscle has been lost in the cat along with the first digit.

A peroneus complex lies posterior to the extensor digitorum longus and takes its origin from the full length of the fibula. The complex is subdivided into three components, which, although they arise from different parts of the fibula, are most distinct at their insertion. In the cat the tendon of the *peroneus longus* passes through a groove on the surface of the lateral malleolus, and then runs through a diagonal groove deep in the sole of the foot, attaching onto metatarsals 2 to 4. In human beings its tendon passes posterior to the lateral malleolus and inserts on the first metatarsal. Its primary action appears to be abduction and eversion of the foot, and it assists in plantar flexion. The tendons of the *peroneus tertius* and *peroneus brevis* pass through a groove on the posterior border of the lateral malleolus in the cat; in human beings the tendon of the tertius goes anterior to the malleolus. That of the tertius continues down the dorsum of the fifth digit, finally uniting with the extensor tendon of the extensor digitorum longus. The much stouter tendon of the peroneus brevis inserts onto the lateral side of the fifth metatarsal, and the tendon of the human muscle also unites with the extensor digitorum longus. These muscles help extend the foot and toes.

CRANIAL TRUNK MUSCLES

The caudal trunk muscles were described before the appendages were dissected, and other trunk muscles were seen during the dissection of the shoulder. Now that the appendages have been examined, it is possible to resume the study of the trunk muscles.

(A) HYPAXIAL MUSCLES

All the trunk muscles that become associated with the pectoral girdle belong to the hypaxial group. They are the *omotransversarius, rhomboideus, rhomboideus capitis,* and *serratus ventralis.* Find them again.

Lay your specimen on its back, reflect the pectoralis, and examine the muscles on the ventrolateral portion of the thoracic wall. The *rectus abdominis* will be seen passing forward to its insertion on the sternum and cranial costal cartilages (Fig. 4–23). Recall that it does not extend so far forward in human beings. The thoracic wall is composed of three layers of muscle comparable to those of the abdominal wall. Observe that the outermost layer, the *external intercostals,* consists of fibers that pass from one rib caudally and ventrally to the next caudal rib. This layer does not extend all the way to the midventral line. Cut through and reflect an external intercostal, and you will find an *internal intercostal.* Its fibers extend cranially and ventrally. The third layer, *transversus thoracis,* is incomplete and found only near the midventral line. To see it, lift up the rectus abdominis, and cut through and reflect the ventral portion of an internal intercostal. The transversus thoracis arises from the dorsal surface of the sternum and is inserted by a number of slips into the costal cartilages. A better view of the muscle will be had when the thorax is opened (Fig. 7–11, p. 134).

In addition to these layers, other muscles are associated with the thoracic wall. The diagonal muscle that arises near the middle of the sternum, and crosses the cranial end of the rectus abdominis to insert on the first rib, is the *rectus thoracis* (Fig. 4–23). Dorsal to its insertion is a fan-shaped complex of muscle which extends between the cervical vertebrae and the ribs. This is the *scalenus.* It arises from the transverse processes of most of the cervical vertebrae and has multiple insertions extending from the first through the ninth ribs. Its insertion is limited to the first two ribs in human beings, and it is subdivided into superior, middle, and interior parts.

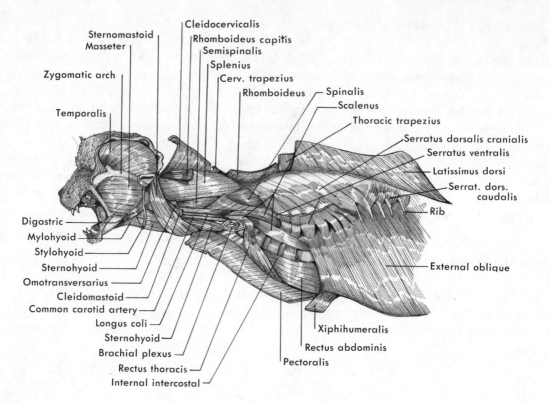

Figure 4–23 Lateral view of the muscles of the neck and thorax after removal of the shoulder and arm.

Turn your specimen on its side, reflect the latissimus dorsi and pull the dorsal margin of the scapula away from the trunk. Medial to the scapula and serratus ventralis, you will see a number of short muscular slips that arise from the thoracolumbar fascia and insert on the dorsal portion of the ribs. These constitute the *serratus dorsalis* (Fig. 4–23). They can be seen better by making a longitudinal incision through the thoracolumbar fascia and reflecting it. The slips can be sorted into a cranial and caudal group. The comparable human muscle is known as the serratus posterior.

All these thoracic muscles, together with the muscular diaphragm and the abdominal muscles, are responsible for respiratory movements. The muscular movements of respiration are very complex, but the major movements during inspiration are a contraction of the diaphragm, which compresses the abdominal viscera, and a forward movement of the ribs through the action of such muscles as the scalenus and the cranial portion of the serratus dorsalis. During expiration, these muscles relax; the diaphragm is pushed cranially by the contraction of abdominal muscles and the pressure of the abdominal viscera; and the ribs move caudally by their elastic recoil and by the action of such muscles as the rectus thoracis and the caudal part of the serratus dorsalis. Electromyographic studies in human beings suggest that the intercostal muscles do not have such a direct role in respiratory movements as formerly believed; their major role appears to be to maintain the ribs a constant distance apart as other forces expand and contract the thoracic cage.

The subvertebral portion of the cranial hypaxial musculature is represented in cat and human beings primarily by the *longus colli* (Fig. 4–23). This muscle appears as a band in the neck lying ventral and medial to the origin of the scalenus. It arises from the bodies of the first six thoracic vertebrae, and as it passes forward it receives other slips of origin from the transverse processes and bodies of the cervical vertebrae. Portions of it insert on the bodies and transverse processes of each of the cervical vertebrae. Its action is flexion and lateral flexion of the neck.

(B) EPAXIAL MUSCLES

The most superficial of the epaxial muscles on the back of the neck is the *splenius* (Fig. 4–23). It is a thin but broad triangular muscular sheet lying deep to the trapezius and the cranial portions of the rhomboideus and serratus dorsalis. The splenius arises from the middorsal line of the neck and passes forward and laterally to insert on the occipital region of the skull (nuchal crest) and transverse process of the atlas. The

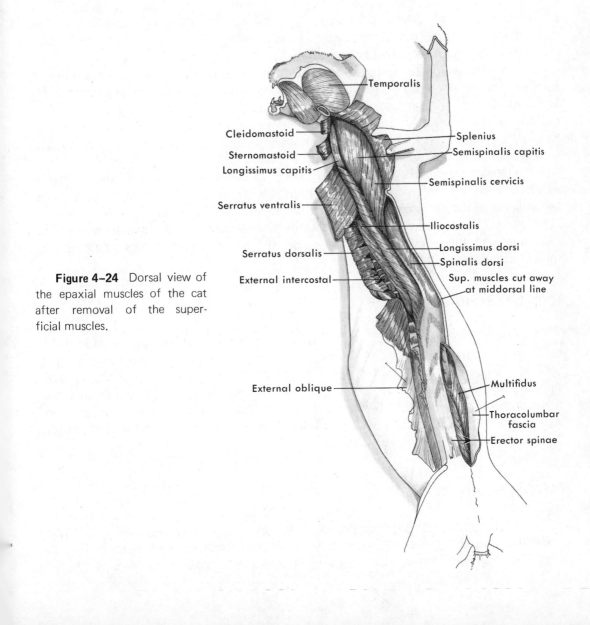

Figure 4–24 Dorsal view of the epaxial muscles of the cat after removal of the superficial muscles.

insertion of the human splenius extends caudad to the fourth cervical vertebra. Each splenius individually acts as a lateral flexor of the head; together they elevate the head (extension).

Much of the thoracolumbar fascia was reflected during the study of the caudal epaxial muscles and the serratus dorsalis. Complete the reflection of this fascia. The epaxial mass should now be well exposed. Find the multifidus and erector spinae caudally (p. 48) and trace them forward.

The *multifidus* can be seen most clearly in the lumbar region (Fig. 4–24). It consists of bundles of muscle fibers that extend from the mamillary processes, transverse processes, and zygapophyses of caudally lying vertebrae to the spinous processes of more cranial ones. Most bundles cross two vertebrae between their origin and insertion. More cranial parts of the multifidus lie deep to the spinalis and will not be seen.

The *spinalis* lies lateral to the spinous processes of the thoracic and the more posterior cervical vertebrae (Fig. 4–24). Most of it arises from the fascia covering the erector spinae, and for this reason the spinalis is sometimes considered to be a division of the erector spinae. Deeper parts arise from the dorsal surface of the vertebrae. The spinalis inserts on the spinous processes of the vertebrae. Fibers of both the multifidus and spinalis extend diagonally craniomedially.

As the *erector spinae* continues forward from its origin on the iliac crest and dorsal surfaces of the more caudal trunk vertebrae, it splits into a *longissimus dorsi,* lying lateral to the spinalis, and a more lateral *iliocostalis.* Fibers of these muscles extend diagonally craniolaterally; those of the iliocostalis insert on the ribs, those of the longissimus dorsi insert chiefly on the transverse processes of thoracic and cervical vertebrae. A *longissiumus capitis* continues the bundle to the head.

The group of muscle bundles lying deep to the splenius, and arising from the vertebrae between the cranial ends of the spinalis and longissimus dorsi, constitute the *semispinalis cervicis et capitis.* They insert on the back of the skull.

Deeper epaxial muscles are not described. Collectively, the epaxial muscles are extensors and lateral flexors of the back, the neck, and the head. Our epaxial musculature is very similar to the cat's.

HYPOBRANCHIAL MUSCLES

The hypobranchial muscles are an axial group located on the ventral side of the neck and throat. All move the larynx, hyoid apparatus, and tongue. The group is utilized in opening the mouth, manipulating food, and swallowing. Complete the skinning of this region as far forward as the chin. Clean away the loose connective tissue and fat. You may turn back, but do not destroy, prominent glands, ducts, blood vessels, and nerves. It is convenient to subdivide the hypobranchial muscles into groups lying caudal and cranial to the hyoid bone.

(A) POSTHYOID MUSCLES

Find the sternomastoid muscles (p. 49) and push them laterally. The thin, midventral band of muscle that covers the windpipe (trachea), and extends from the cranial end of the sternum to the hyoid, is the *sternohyoid* (Figs. 4–6, 4–25, and

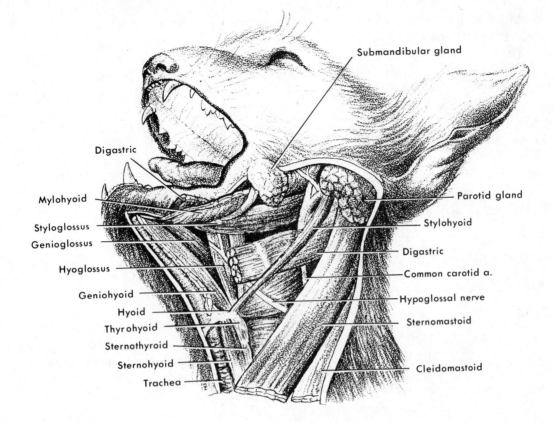

Submandibular gland

Digastric

Mylohyoid

Styloglossus

Genioglossus

Hyoglossus

Geniohyoid

Hyoid

Thyrohyoid

Sternothyroid

Sternohyoid

Trachea

Parotid gland

Stylohyoid

Digastric

Common carotid a.

Hypoglossal nerve

Sternomastoid

Cleidomastoid

Figure 4–25 Lateroventral view of the musculature in the cranial part of the neck and the floor of the mouth in the cat.

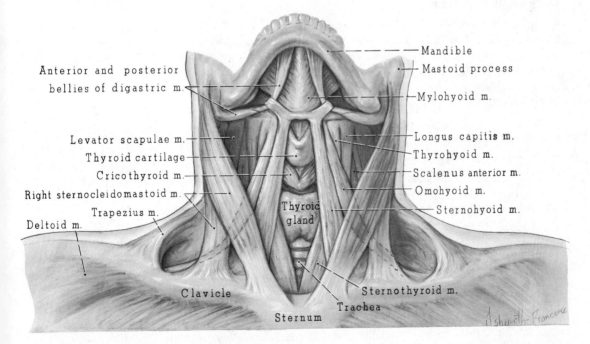

Anterior and posterior
bellies of digastric m.

Levator scapulae m.

Thyroid cartilage

Cricothyroid m.

Right sternocleidomastoid m.

Trapezius m.

Deltoid m.

Mandible

Mastoid process

Mylohyoid m.

Longus capitis m.

Thyrohyoid m.

Scalenus anterior m.

Omohyoid m.

Sternohyoid m.

Thyroid gland

Clavicle

Sternum

Sternothyroid m.

Trachea

Figure 4–26 Ventral or anterior view of the muscles in the neck of a human being. (From Jacob and Francone: Structure and Function in Man. W.B. Saunders Company.)

4–26). Actually, there is a pair of sternohyoids, but they are generally fused in the cat. Their origin is the sternum; their insertion, the hyoid.

Carefully separate the sternohyoid from another band of muscle that lies dorsal and lateral to it. This band, the *sternothyroid,* has a similar origin but passes forward to insert on the thyroid cartilage of the larynx. The larynx, or Adam's apple, lies caudal to the hyoid. Thus, the sternothyroid is not so long a muscle as the sternohyoid. A short band of muscle, the *thyrohyoid,* lies on the lateral surface of the larynx. It arises at the point of insertion of the sternothyroid and passes forward to insert on the hyoid. Sternothyroid and thyrohyoid appear as one band unless their attachments on the larynx are carefully exposed.

Our posthyoid muscles are similar to those of the cat except that we also have an omohyoid extending from the scapula to the hyoid (Fig. 4–26).

(B) PREHYOID MUSCLES

Two branchiomeric muscles must be studied and reflected in order to expose the prehyoid muscles. Note the stout band of muscle that is attached to the ventral border of the mandible. This is the *digastric.* It has a fleshy origin from the paracondyloid and mastoid processes of the skull, inserts along the ventral border of the lower jaw, and acts to open the jaw. Its name derives from the condition in human beings in which it is divided by a central tendon into two bellies. Sometimes a connective tissue tendinous intersection can be seen in this position in the cat. Disconnect its insertion from the jaw on one side, and partially reflect the muscle. Do not break the narrow ribbon of muscle (stylohyoid, Figs. 4–6 and 4–25) that extends across it. The sheet of more or less transverse fibers that lies between and deep to the insertions of the digastric muscles of opposite sides of the body is the *mylohyoid.* It arises from the mandible and inserts on a median tendinous intersection and on the hyoid. It acts in both cat and human beings to raise the floor of the mouth and to pull the hyoid forward. Make a longitudinal incision through the muscle and reflect it on one side; it is not very thick.

The longitudinal muscles that lie deep to the mylohyoid constitute the prehyoid portion of the hypobranchial musculature. It may be necessary to cut through the mandibular symphysis and spread the two halves of the lower jaw apart to see these muscles clearly. The midventral band of muscle (really a pair of muscles that have united) is the *geniohyoid* (Fig. 4–25). It arises from the front of the mandible and inserts on the hyoid. Cut across the geniohyoid and reflect its ends. The muscle that arises from the hyoid lateral, and deep, to the insertion of the geniohyoid is the *hyoglossus.* It passes forward into the tongue. Pull the tip of the tongue, and you will note that the muscle is moved. The band of muscle that arises from the chin deep to the origin of the geniohyoid is the *genioglossus.* It passes caudally into the tongue, lying medial to the hyoglossus. The band of muscle that arises from the mastoid process at the base of the skull and passes forward into the tongue is the *styloglossus.* It lies lateral to the rostral portion of the hyoglossus. The glossus muscles, together with intrinsic muscle fibers within the tongue *(lingualis proprius),* form the substance of the tongue and manipulate this organ. Human prehyoid muscles are the same.

BRANCHIOMERIC MUSCLES

(A) MANDIBULAR MUSCLES

The *mylohyoid* and the cranioventral half of the *digastric,* which were seen during the dissection of the hypobranchial muscles, are branchiomeric muscles of the mandibular arch. To see the other mandibular arch muscles in the cat, skin the top of the head and the cheek region on one side. The auricle should also be cut off. The platysma, facial muscles (both belonging to the hyoid group), and loose connective tissue must be removed, but be careful not to injure glands, nerves, and blood vessels in this region. Special care should be exercised in skinning and cleaning the cheek, for the duct of the parotid salivary gland crosses the cheek just beneath the facial muscles (Fig. 4–9). Find the zygomatic arch. The powerful muscle that lies ventral to the arch is the *masseter.* It arises from the arch and inserts in the masseteric fossa and adjacent parts of the mandible (Fig. 4–9).

Another mandibular muscle, *temporalis,* lies dorsal to the zygomatic arch. In the cat it is a sizable muscle, and fills the large temporal fenestra (Fig. 4–23). It arises primarily from the surface of the cranium, but some fibers spring from the top of the zygomatic arch. It passes deep to the zygomatic arch and inserts on the coronoid process of the mandible.

Other mandibular muscles are impractical to dissect at this stage. Two pterygoid muscles pass from the base of the skull to the medial side of the ramus of the mandible. They will be seen when the mouth and the pharynx are opened (p. 130). A tensor tympani lies in the middle ear, where it attaches to the malleus, and a tensor veli palatini extends from the base of the skull into the soft palate. The tensor tympani is described in more detail in connection with the ear (p. 91).

The masseter, temporalis and pterygoids close the jaws. The digastric is the major muscle involved in opening the jaws, but the mylohyoid and hypobranchial muscles can assist under certain circumstances. The mylohyoid also compresses the floor of the mouth and elevates the hyoid. The functions of the tensor tympani and tensor veli palatini are implied by their names.

Jaw mechanics vary among mammals (Fig. 4–27). The jaw joint of a cat, as is characteristic of carnivores, is in line with the tooth row, so that the upper and lower jaws come together in the manner of a pair of scissors. The shape of the condyle and mandibular fossa is such that only a hinge action is permitted. However, the lower jaw as a whole can move slightly to the left or right side so that the carnassial teeth can engage more intimately on one side or the other. In the rabbit, as is characteristic of gnawing and herbivorous mammals, the jaw joint is situated well dorsal to the tooth row so that all of the teeth of the upper and lower jaw come together simultaneously. The shape of the condyle and mandibular fossa permits the lower jaw to move back and forth and from side to side, actions which occur in gnawing and grinding.

The adductor muscles of the jaws must exert a strong force for closing the jaws, and also balance forces at the jaw joint so that there is little tendency for the jaws to become disarticulated. Both masseter and temporal muscles are large in carnivores, but a low condyle and a high coronoid process give the temporal muscle a longer moment arm (the perpendicular distance between the line of action of the muscle and the jaw joint), and hence a greater mechanical advantage, than the masseter. Both exert strong forces for closing the jaws. The direction of pull of the temporal muscle also resists any forward pull on the canine that may occur when a carnivore is seizing prey, and the pull of the masseter helps to resist any tendency for the jaw condyle to slip ventrally and caudally out of the mandibular fossa, which could occur if the temporal muscle alone were used in cutting up the food. The pterygoids of carnivores are modest-sized muscles whose actions are similar to those of the masseters.

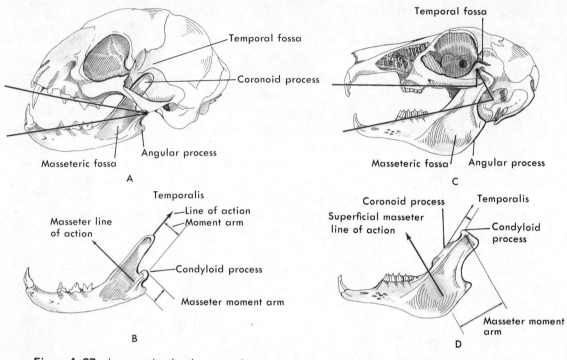

Figure 4–27 Jaw mechanics in a cat (*A* and *B*) and rabbit (*C* and *D*). *A* and *C*, jaw closure; *B* and *D*, lines of action and moment arms of the masseter and temporalis muscles.

The masseter of herbivores is a very large muscle divided into superficial and deep layers that are oriented nearly at right angles to each other. The superficial layer provides a particularly strong force for closing the jaws because it has a very long moment arm. Its tendency to pull the lower jaw forward out of the mandibular fossa is opposed by the deep layer. These two layers, and a large pterygoid, provide the forces needed to close the jaws, move them back and forth and from side to side in a grinding action, and balance forces at the jaw joint. The temporal muscle is not an important muscle in herbivores, and it, together with its points of attachment on the skull and jaws, is small.

We are omnivores, and our jaw and jaw muscles are intermediate in many ways between those of a carnivore and those of a herbivore. The jaw joint lies well superior to the tooth row, as in a herbivore, but the joint is of the hinge type. Although relatively not so large as in a rabbit, the angular region of our lower jaw is well developed, and this increases the area for the insertion of a powerful masseter. The temporalis is also a well-developed muscle, and it has a good lever arm because the coronoid process on which the muscle inserts lies well in front of the condyloid process.

(B) HYOID MUSCLES

The major hyoid arch muscles are the *platysma, facial muscles,* caudal half of the *digastric,* and the *stylohyoid.* The first three have been seen. The stylohyoid is a small ribbon of muscle that lies lateral to the caudal portion of the digastric (Fig. 4–25). It arises from the stylohyal bone, inserts on the body of the hyoid, and acts on the hyoid. We have a similar muscle.

Another hyoid arch muscle, which is not feasible to find at this time, is the stapedius. It lies within the middle ear and acts on the stapes. The muscle is described more fully on page 91.

(C) POSTERIOR BRANCHIOMERIC MUSCLES

Much of the caudal branchiomeric musculature is lost during the course of evolution. But some becomes associated with the pectoral girdle as the *trapezius* and *sternocleidomastoid* complexes. These were described in connection with the shoulder region. Some of the remaining caudal branchiomeric musculature forms the *intrinsic muscles of the larynx* (e.g., *thyroarytenoid, cricoarytenoid, cricothyroid;* Fig. 7–10, p. 133), and some contributes to the wall of the pharynx. Certain of the intrinsic muscles of the larynx can be seen on the ventral surface of the larynx deep to the cranial end of the sternohyoid.

Chapter Five

THE SENSE ORGANS

Although the sense organs and the nervous system integrate the activities of all parts of the body, they may appropriately be considered at this time, for the most conspicuous effector organs are the muscles described in the previous chapter. The sense organs, the central nervous system, and the basic pattern of distribution of the peripheral nerves are the topic of this and the following chapter. If separate sheep brains cannot be provided for study, this unit of work should be postponed until the end of the course.

IRRITABILITY

Irritability, that is, the ability to receive sensations and respond to stimuli, is a basic property of protoplasm. All the aspects of irritability are combined in the individual cells of the more primitive organisms, but in higher animals there is a division of labor. Certain cells receive, others transmit, and still others respond to the stimuli. The receptive cells are the *receptors* within the sense organs; the transmitting cells are the neurons of the *nervous system;* and the responding cells are the *effectors* (muscles, glands, cilia, etc.).

CLASSIFICATION OF SENSE ORGANS

The stimuli causing the sensation of pain may be received by free nerve endings, but stimuli producing other sensations in vertebrates are received by very specific sense organs. These range from microscopic sensory corpuscles to such large and complex organs as the eye. This array of sense organs, and the sensory nerves which lead from them, may be divided into two groups—somatic and visceral. *Somatic sensory organs* lie in the "outer tube" of the body, that is, in the skin (exteroceptors of the physiologist) and in the somatic muscles (proprioceptors of the physiologist). Occasionally, proprioceptive organs, which are the organs of muscle sense, are found in branchiomeric muscles. *Visceral sensory organs* (interoceptors) are associated with the "inner tube" of the body, i.e., the viscera. Only those sense organs that can be seen grossly will be considered.

THE EYEBALL AND ASSOCIATED STRUCTURES

The eyes are somatic sensory organs. Two types are found in vertebrates: the conventional pair of *lateral eyes* and a *median eye* on the top of the head. Median eyes, either *pineal* or *parietal,* develop as outgrowths from the diencephalic region of the brain (Fig. 5–1, *A* and *B*). They are not visual but simply light receptive. As one ascends the evolutionary scale, the parietal eye becomes adapted

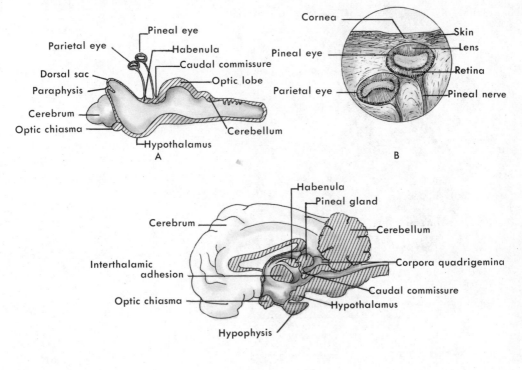

Figure 5–1 Diagrams of the pineal-parietal complex. *A*, relationship of the median eyes to the brain in a lamprey; *B*, enlargement of the median eyes of a lamprey; *C*, the pineal gland of a sheep and its relation to the brain.

primarily for light reception. It is well developed in many reptiles, in which it detects the amount of solar radiation and initiates behavioral responses by which the animal regulates its body temperature. The parietal eye is lost in birds and mammals, which have evolved physiological mechanisms of thermoregulation. The pineal eye, or epiphysis, tends to lose its photoreceptive function and become glandular, forming in mammals a well-developed pineal gland (Fig. 5–1, *C*), which will be seen when the brain is studied. Secretion of the pineal gland is affected by light received by the parietal eye or in other ways.

The lateral eye will be considered at this time. Although the eyeball differs in its method of accommodation and in its adaptive details, its basic anatomy is much the same in all vertebrates. Evolutionary tendencies, however, can be seen in associated structures. For example, the surface of the fish eye is bathed in water which keeps it moist and clean. Tetrapods have evolved movable eyelids of various types and tear glands and ducts that protect, cleanse, and moisten the eye surface.

The accessory structures associated with the eyeball should be dissected on the side of your specimen that was used for the study of the muscles. Movable *upper* and *lower eyelids (palpebrae)* are present. The slitlike opening between them is called the *palpebral fissure.* The corners of the eye, where the lids unite, are the *ocular angles.* Cut through the lateral ocular angle and pull the lids apart. The *nictitating membrane* (semilunar fold) can now be seen clearly. It is attached at the medial ocular angle, but its lateral edge can spread over the surface of the eye if the eyeball is retracted slightly. Our nictitating membrane is reduced to a vestigial semilunar fold *(plica semilunaris)* that can be seen covering the medial corner of the eye (Fig. 5–2). Examine the edge of each lid three or four millimeters from the medial ocular angle with a hand lens. If you

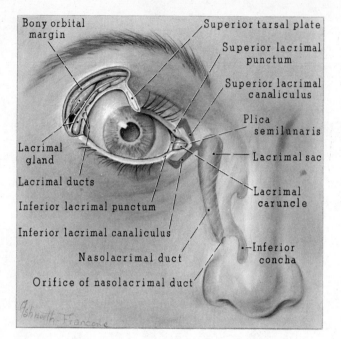

Bony orbital
margin
Superior tarsal plate
Superior lacrimal
punctum
Superior lacrimal
canaliculus
Plica
semilunaris
Lacrimal
gland
Lacrimal sac
Lacrimal ducts
Lacrimal
caruncle
Inferior lacrimal punctum
Inferior lacrimal canaliculus
Nasolacrimal duct
Inferior
concha
Orifice of nasolacrimal duct

Figure 5–2 Anterior view of the lacrimal apparatus of a human being. (From Jacob and Francone: Structure and Function in Man. W.B. Saunders Company.)

are fortunate, you will see on each a minute opening *(lacrimal punctum)* that leads into a small lacrimal canal. If you cannot find them in your specimen, look for one on the human eye by pulling down the lower lid and examining its edge near the most medial eyelash.

Cut off the upper and lower lids, leaving a bit of skin around the medial ocular angle. As you remove the lids, note that a thin, transparent membrane (the *conjunctiva*) lines the underside of the lids and reflects over the *cornea* of the eyeball. If the cornea is not too opaque, the *pupil* and *iris* can be seen. A facial muscle, the *orbicularis oculi,* encircles the eyelids, and will be cut off with them (Fig. 4–3, p. 44). It closes the lids.

Free the eyeball and associated glands from the bony rims of the orbit by picking away connective tissue. Do not dissect deeply in the region of the medial ocular angle, and try not to destroy a loop of connective tissue attached to the anterodorsal wall of the orbit. One of the ocular muscles passes through this loop. Using bone scissors, cut away the zygomatic arch beneath the eye and the postorbital processes. Push the eye anteriorly. The dark glandular mass that lies on the posterodorsal surface of the eyeball is the *lacrimal gland* (Fig. 5–2). Its secretions, the tears, enter near the lateral ocular angle, bathe the surface of the cornea, and pass into one of the small lacrimal canals through a *lacrimal puncta.* The lacrimal canals from each lacrimal punctum unite to form a nasolacrimal duct that enters the nose through a canal in the *lacrimal bone.* Find the lacrimal canal on a skull (p. 25). You may be able to find the nasolacrimal duct by dissecting ventral and anterior to the medial ocular angle. First find the anterior border of the orbit and the position of the lacrimal canal. Do not injure a muscle that is attached to the orbital wall near the lacrimal bone.

The nictitating membrane of the cat contains a small *gland of the nictitating membrane* (harderian gland), which may be found by removing and looking on the medial surface of the base of the nictitating membrane. It supplements the secretions

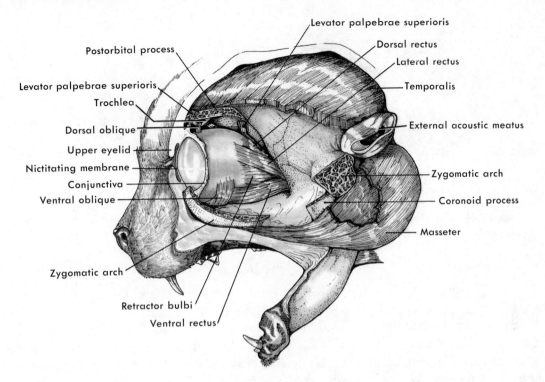

Figure 5–3   Lateral view of a dissection of the orbit of a cat to show the extrinsic ocular muscles.

of the lacrimal gland. We lack this gland, but in other respects our lacrimal apparatus is similar to the cat's.

Push the eyeball dorsally and carefully remove fat and connective tissue from the floor of the orbit. A small salivary gland (the *zygomatic gland*) lies on the floor of the orbit in line with, and just posterior to, the upper tooth row. We do not have this gland.

Remove the glands and connective tissue from around the eye and expose the extrinsic muscles of the eyeball (Fig. 5–3). A *levator palpebrae superioris* arises from the medial wall of the orbit dorsal to the optic foramen and inserts on the upper eyelid, which it raises. Its lateral end may be found unattached, for its insertion may have been cut in removing the eyelids. The rest of these muscles move the entire eyeball. Two oblique muscles pass to the lateral wall of the eyeball. The *dorsal* (superior) *oblique* arises from the wall of the orbit slightly anterior to the optic foramen and goes through a connective tissue pulley (the *trochlea*), which is attached to the anterodorsal wall of the orbit, before inserting on the eyeball. The *ventral* (inferior) *oblique* arises from the maxillary bone and attaches to the lateral surface of the eyeball ventral to the insertion of the superior oblique. Four recti arise from the margins of the optic foramen and pass to insert on a line around the eyeball several millimeters deep to the margin of the cornea. A *dorsal* (superior) *rectus* attaches on the dorsal surface of the eyeball; a *ventral* (inferior) *rectus,* on the ventral surface; a *medial rectus,* on the medial surface; and a *lateral rectus,* on the lateral or posterior surface. A *retractor bulbi,* which can be divided into four parts, passes to the eye deep to the recti. We do not have a retractor bulbi, but our other extrinsic ocular muscles are the same as in the cat.

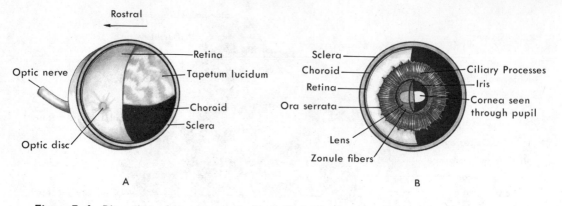

Figure 5–4 Dissection of a cow eye. *A*, Posterior half; *B*, anterior half. A portion of the retina has been removed in each half to show the vascular tunic.

The structure of the *eyeball (bulbus oculi)* can best be studied from preserved specimens of cow or sheep eyes, although the cat's may be used. The following description is based on the cow eye (Fig. 5–4); a human eye is essentially the same. Carefully clean up the surface of the eyeball by removing the extrinsic ocular muscles and associated fat. Notice that the optic nerve attaches somewhat excentrically, toward the anteroventral portion of the eyeball. Leave a stump of it attached. Some conjunctiva will adhere to the surface of the cornea.

Open the eyeball by carefully cutting a small window through its dorsal surface. Observe that its wall consists of three layers: (1) an outer *fibrous tunic;* (2) a middle dark layer, the *vascular tunic;* and (3) the inner, whitish *retina.* The fibrous tunic is a dense, supporting connective tissue. Approximately the medial two-thirds of it form the opaque *sclera;* the lateral one-third, the transparent *cornea* through which light enters the eye. The vascular tunic is rich in blood vessels; most of this layer is a *choroid* that lies behind the retina and helps to nourish it. The portion of the retina which you see is the nervous layer which contains the receptive rods and cones on its choroid surface. Embryonically there is a pigmented layer to the retina, but this becomes associated with the choroid in the adult. The pigment reduces light reflections within the eye.

With a pair of fine scissors, extend a cut from the window which you made completely around the equator of the eyeball, thereby separating the eyeball into an anterior half containing the lens and cornea and a posterior half containing most of the retina. Cut through all the layers. The jelly-like mass filling the eyeball between the lens and retina is the *vitreous body,* a medium which helps to support the lens and also helps to refract light entering the eye. The space within which it lies is the *chamber of the vitreous body.* Keep the vitreous body with the anterior half of the eyeball. Submerge both halves in a dish of water.

Examine the posterior half (Fig. 5–4, *A*). The retina has probably become partly detached from the choroid but can be floated back into its normal position. Note the round spot (*optic disc*) at which the optic nerve attaches to the retina. This region is devoid of rods and cones, hence it is often called the blind spot. Remove the retina and observe that an extensive section of the choroid dorsal to the optic disc is quite iridescent. This is the *tapetum lucidum,* an area which reflects some of the light passing through the retina back onto the rods and cones and, hence, facilitates the animal seeing in dim light. It is formed by a compact layer of endothelial cells in the choroid. Many mammals, including human beings, lack a tapetum.

Carefully remove the remains of the vitreous body from the anterior half of the eyeball; notice that the white, nervous layer of the retina does not extend far into this half of the eyeball. The line of demarcation between it and the dark choroid (plus embryonic pigment layer of retina) is the *ora serrata* (Fig. 5–4, *B*). The portion of the choroid coat which you see extending from the ora serrata toward the *lens* is the *ciliary body.* The portion of it next to the lens has a pleated appearance, and the individual folds are the *ciliary processes.* While observing the area with a dissecting microscope, carefully stretch the region between the lens and the ciliary processes. Many delicate *zonule fibers* will be seen passing from the ciliary body to the equator of the lens.

Intraocular pressures cause the wall of the eyeball to bulge outward, and this force is transmitted via the zonule fibers to the elastic lens, which is consequently under tension and somewhat flattened. Under these circumstances the lens has its minimum thickness and refractive powers, so the eye is focused on distant objects. It should be pointed out that the major refraction of light is caused by the cornea; the role of the lens is more analogous to the fine adjustment of a microscope. Accommodation for a close object requires a greater bending of light rays. Muscle fibers in the base of the ciliary body contract and bring the base of the ciliary body a bit closer to the lens. This releases the tension on the zonule fibers and permits the lens to bulge and increase its thickness.

Carefully remove the lens and notice that the choroid coat continues in front of it to form the *iris.* The *pupil,* of course, is the opening through the iris. Its diameter, and the amount of light it permits to pass, are regulated by circular and radial muscle fibers within the iris. The space between the lens and iris is the *posterior chamber;* that between the iris and cornea, the *anterior chamber.* Both are filled with a watery *aqueous humor* secreted by the ciliary processes. This liquid maintains the intraocular pressure. Excess liquid is drained off by a microscopic *scleral venous sinus* (canal of Schlemm), which encircles the eye between the base of the cornea and the iris. If one makes a vertical cut through the anterior half of the eyeball and examines it under a dissecting scope, this canal and the ciliary muscles can sometimes be seen.

THE NOSE

In fishes, the nose typically consists of a pair of sacs, each of which connects to the surface by a pair of openings, external nostrils *(nares),* through which water carrying odoriferous particles circulates. In fishes ancestral to tetrapods, and in tetrapods, each olfactory sac, now usually called a *nasal cavity,* has but one naris, but each opens into the mouth through an internal nostril *(choana).* Thus the nose serves as an air passage as well as retaining its orginal olfactory function. In the evolution through tetrapods, the olfactory and respiratory roles of the nasal passages become segregated to some extent, the olfactory epithelium becoming restricted for the most part to the dorsal parts of the cavities. But in many tetrapods a part of the olfactory epithelium remains in the ventral part of the passage where it forms the *vomeronasal organ* (Jacobson's organ). This organ apparently detects the odors of food within the mouth, and it sometimes has a separate connection with the mouth (p. 129). In mammals the respiratory passages are prolonged through the evolution of a bony hard palate (p. 17) and a fleshy soft palate.

In order for a substance to stimulate the olfactory epithelium it must be in solution. This is no problem for the fish, but it is for a terrestrial animal. Tetrapods have met the problem by the evolution of glands whose secretions keep the epithelium moist. The secretions of these glands, and the mucosa of the nasal passages as a whole also condition the air that passes to the lungs by moistening, cleansing, and, in birds and mammals, warming it. In birds and mammals the mucosal surface is increased through the evolution of *conchae* or turbinate bones (p. 24).

The receptive olfactory cells of the nasal organ are of a unique type in that the cells not only receive the stimuli but also have long processes that conduct the impulses back to the brain. This type of cell, which is believed to be very primitive, is called a *neurosensory cell.* Although related to the visceral sensation of taste, olfaction is considered to be somatic sensory.

Study the nose on sagittal sections of the head cut in such a way that one half shows the nasal septum, and the other the inside of the nasal cavity. The nose should be studied from demonstrated preparations, unless this unit of work has been postponed to the end of the course, or unless heads from specimens of a previous year's class have been saved for the purpose. This work should also be correlated with the description of the sagittal section of the skull on page 22.

The *nares,* which are close together in mammals, lead into paired *nasal cavities.* The nasal cavities occupy the area of the skull anterior to the cribriform plate of the ethmoid bone and dorsal to the bony, hard palate. On the larger section, it can be seen that they are completely separated from each other by a *nasal septum.* The ventral portion of the septum is formed by the *vomer bone,* the posterodorsal portion by the *perpendicular plate of the ethmoid,* and the rest by *cartilage.*

On the smaller section, it can be seen that each nasal cavity is filled to a large extent by three folded *conchae* or turbinate bones (Figs. 2–9 and 7–7). These are, of course, covered with the nasal mucosa. The *ventral* (inferior) *concha* (maxilloturbinate) is represented by a simple fold that extends from the dorsal edge of the naris posteriorly and ventrally to about the middle of the hard palate. The nasolacrimal duct from the eye enters lateral to the inferior concha. It is best seen on a skull in which this concha has been removed. The *dorsal* (superior) *concha* (nasoturbinate) is represented by a single longitudinal fold in the dorsal part of the nasal cavity lying deep to the median, perpendicular plate of the nasal bone. The area between and posterior to these two conchae is filled by the highly folded *middle concha* (ethmoturbinate). The spaces among the folds of the middle concha are referred to as the *ethmoid cells.*

Air passages lead from the external naris between the conchae. The most prominent of these is the *ventral* (inferior) *meatus,* which lies between the ventral and middle conchae on the one hand and the hard palate on the other. It opens by the *choana* into the nasopharynx. The choanae are located at the posterior border of the hard palate. The nasopharynx is separated from the oral pharynx by the fleshy soft palate. A *dorsal* (superior) *meatus* extends from the dorsal part of the naris to the caudal portion of the middle concha. A *common meatus* lies between the nasal septum and the conchae, and other air passages lie between the complex scrolls of the conchae. Sinuses in the frontal and ethmoid bones connect with certain of the air passages. Most of the olfactory epithelium is limited to the more posterior ethmoid cells. *Olfactory nerves* may be seen, with a hand lens, passing through the cribriform foramina of the cribriform plate to the olfactory bulb of the brain. Our conchae are not so complexly folded and the air passages between them more evident (Fig. 5–5).

Paired *vomeronasal organs* are present in the cat but absent in human beings. The entrance to one can be seen on the roof of the mouth just posterior to the first incisor tooth (Fig. 7–8). From here an *incisive duct* leads through the palatine fissure (p. 25) to the organ. The vomeronasal organ can be found by carefully dissecting away the

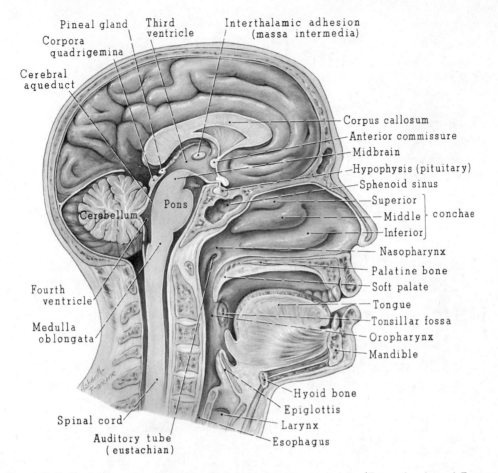

Figure 5-5 Sagittal section through the head of a human being. (From Jacob and Francone: Structure and Function in Man. W.B. Saunders Company.)

anteroventral portion of the nasal septum. It appears as a cul-de-sac with a cartilaginous wall lying on the hard palate and against the nasal septum. It extends between one-fourth and one-half inch caudad to the incisor teeth (Fig. 7-7).

THE EAR

The inner ear of vertebrates consists of a series of thin-walled canals and sacs filled with a fluid known as the *endolymph.* These canals and sacs are collectively called the *membranous labyrinth,* and they are embedded within a series of parallel canals and chambers within the otic capsule known as the *osseous labyrinth.* The membranous labyrinth and osseous labyrinth are separated from each other by spaces filled with fluid and crisscrossed by minute strands of connective tissue. This fluid is the *perilymph.*

Only an internal ear is present in fishes (Fig. 5-6). As in high vertebrates, it is an organ of equilibrium, and in many fishes the ventral part of it has also been shown to be sensitive to sound waves that reach it by passing from the water through the adjacent tissues. In terrestrial vertebrates, a part of the inner ear is definitely specialized for the reception of sound waves.

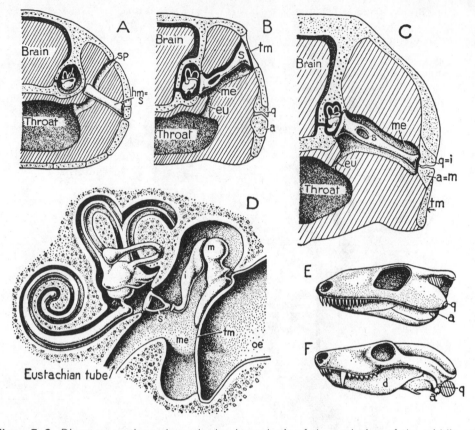

Figure 5–6 Diagrams to show the orthodox hypothesis of the evolution of the middle ear and auditory ossicles. The cross sections are through the otic region of *A*, a fish; *B*, a primitive amphibian; *C*, a primitive reptile; and *D*, a mammal. The lateral views are of *E*, a primitive amphibian; and *F*, a mammal-like reptile. They show in particular the shift of the eardrum (hatched) from the otic notch to a point behind the jaw articulation. Abbreviations: *a*, articular; *d*, dentary; *eu*, auditory or eustachian tube; *hm*, hyomandibular; *i*, incus; *m*, malleus; *me*, middle ear, or tympanic, cavity; *oe*, outer ear cavity (external acoustic meatus); *q*, quadrate; *s*, stapes; *sp*, spiracle; *tm*, tympanum, or eardrum. (From Romer: Man and the Vertebrates. University of Chicago Press.)

Whereas sound waves pass easily from water into the tissues of a fish, which are mostly water, they do not pass easily from air into water. Tetrapods sensitive to airborne sound vibrations have evolved a special mechanism that receives such vibrations and increases their pressure amplitude sufficiently to overcome the inertia in the liquids of the inner ear. In most living amphibians and reptiles, sound waves impinge upon the ear drum, or *tympanum,* located on the body surface or at the base of a canal, the *external acoustic meatus.* They are transmitted across a *tympanic cavity* (homologous to the first gill pouch, or spiracle, of certain fishes) by the *stapes* (homologous to the hyomandibular, or dorsal part of the hyoid arch). The foot plate of the stapes fits into a *fenestra vestibuli,* or oval window, on the side of the otic capsule, and a specialized part of the perilymph transmits vibrations from there to the receptive part of the membranous labyrinth. The difference in size between the large tympanum and small fenestra ovalis increases the pressure amplitude. Vibrations are finally released from the inner ear through a *fenestra cochlea,* or round window.

Early amphibians and reptiles probably had an ear of this type (Fig. 5–6), which has been retained in most of their descendants and has been further elaborated in mammals. The sound-detecting portion of the mammalian inner ear is a long, spiral cochlea; two additional ossicles, *malleus* and *incus*, appear in the middle ear as a corollary of a new jaw joint and the inward movement of the articular and quadrate (the portions of the mandibular arch bearing the jaw joint, p. 17); and an *auricle* develops about the entrance of the external auditory meatus.

The external ear can be seen easily on your specimen. It consists of the external ear flap, the *auricle,* and a canal with a cartilaginous wall leading inward to the skull *(external acoustic meatus).* The *tympanum* lies at the base of the meatus between the external and middle ear. It can be found on the side of the head on which the auricle was removed by cutting away as much of the external acoustic meatus as possible and shining a light into the remainder. Note that the tympanum is set at an angle, its anterior portion extending more medially than its posterior portion. The opaque line seen through the membrane is the handle of the malleus.

The rest of the ear is difficult to dissect and should be observed on demonstration preparations. These can be prepared from the sagittal sections of the head used for the study of the nose. Remove the muscles and other tissue from around the tympanic bulla except at its anteromedial corner. The middle ear, or *tympanic cavity,* lies within the bulla and opens into the nasopharynx by the *auditory tube* (eustachian tube). The opening of the tube appears as a slit in the lateral wall of the nasopharynx (Fig. 7-7). Careful dissection between this slit and the anteromedial corner of the bulla will reveal the tube, part of whose wall is cartilaginous and part bony.

The rest of the dissection will be easier to do if the specimen is first decalcified by placing it in a weak solution (0.06) of nitric acid for a few days. Using bone scissors, break away the posteromedial portion of the bulla and also the mastoid and paracondyloid processes and adjacent parts of the nuchal crest. This exposes the posteromedial chamber of the tympanic cavity. Note that it is largely separated from a smaller anterolateral chamber by a more or less vertical plate of bone. A hole through the dorsolateral portion of this plate passes between the two chambers of the middle ear cavity. The *fenestra cochlea,* or round window, can be seen through this hole. Carefully break away all of this plate of bone and open up the anterolateral chamber. The handle of the *malleus* can be seen on the inside of the tympanum, and it will be noted that the fenestra cochlea is situated on a round promontory of bone. A fingerlike process of cartilage extends from the posterolateral wall of the tympanic cavity between these two structures. The tiny nerve that runs along it and leaves its tip is the *chorda tympani,* a branch of the facial nerve going to the taste buds of the tongue and certain salivary glands. Break away bone from the anteromedial corner of the bulla and find the entrance of the auditory tube. The other auditory ossicles *(incus* and *stapes),* and the *fenestra vestibuli,* or oval window, in which the stapes fits, lie dorsal to the fenestra cochlea. To see them, one must remove a piece of bone posterior and dorsal to the external acoustic meatus without injuring the plate of bone supporting the tympanic membrane. You will also notice two small muscles passing to certain of the ossicles. A *stapedius* arises from the medial wall of the tympanic cavity caudad to the fenestra cochlea and inserts on the stapes. A *tensor tympani* arises from the medial wall anterior to the fenestra vestibuli and inserts on the malleus. These are branchiomeric muscles derived respectively from the musculature of the hyoid and mandibular arches and, hence, innervated by the facial and trigeminal nerves. They adjust the auditory ossicles and tympanum to the intensity of the sound waves. Their contraction, for example, reduces the amplitude of the vibration of the ossicles and protects the delicate structures of the inner ear from violent movements resulting from loud noises. In this respect, they are analogous to the muscles in the iris of the eye, which adjust the eye to light intensity. Our tympanic cavity is not divided into two chambers, but in other respects our auditory apparatus is essentially the same as the cat's.

The inner ear lies within the petrous portion of the temporal bone. Portions of it may be noted by removing the brain and chipping away pieces of the petrous, but it cannot be dissected satisfactorily by this method. The *internal acoustic meatus* for the vestibulocochlear and facial nerves lies in the cranial cavity on the posteromedial surface of the petrosal.

Chapter Six

THE NERVOUS SYSTEM

As explained in the introduction to the chapter on sense organs, the nervous system integrates, or coordinates, the actions of the various parts of the body, so that all will function in harmony with one another and with the external environment. However, it should be pointed out that the nervous system is not the only system involved in body coordination. The endocrine glands also play an important role. Nervous coordination is implemented by nerve impulses that travel along the nerve cells or neurons. It is characterized by being rapid and specific. Endocrine coordination, on the other hand, is by way of the secretions of hormones that are carried by the circulatory system. It tends to be slower, and often one hormone may affect many different parts of the body. Since the endocrine glands are widely scattered, they will not be discussed in one place but as they are observed. The nervous system will be considered at this time, as it is the last of the group of organ systems that deals directly with the general function of body support and movement.

NEURONS

The morphological and functional unit of the nervous system is the *neuron*, just as the muscle cell is of the muscular system. A typical neuron has a *cell body*, which includes the nucleus and most of the cell's cytoplasmic organelles. Extending from the cell body are cytoplasmic processes: functionally, *dendrites* carry information to the cell body, and an *axon* carries information away from the cell body. In most cases the axon is by far the longer process; in large vertebrates it may extend many feet. The one exception is in the case of primary sensory neurons carrying information toward the spinal cord or brain. In this case the dendrite is very long, extending from the skin or sense organ to the cell body which lies just outside the brain or cord; the axon may be much shorter. The term *nerve fiber* refers to a long process, usually an axon. Protective cells surround all nerve fibers, and in many cases contain a fatty substance, *myelin*, which forms many layers of wrapping around the fibers. Aggregates of nerve cell bodies within the brain are known as *nuclei;* aggregates of cell bodies outside the brain and spinal cord, as *ganglia. Tracts* are groups of fibers running together within the brain and cord. *Nerves* are groups of fibers outside the brain and cord.

DIVISIONS OF THE NERVOUS SYSTEM

Grossly, the nervous system can be divided into a central and a peripheral portion (Fig. 6–1). The *central nervous system* consists of the *brain* and *spinal cord*. Both cord and brain are hollow, for they contain a *central canal*, which expands to form *ventricles* in certain regions of the brain. The distribution of the neurons within the central nervous system is such that we can speak of *gray* and

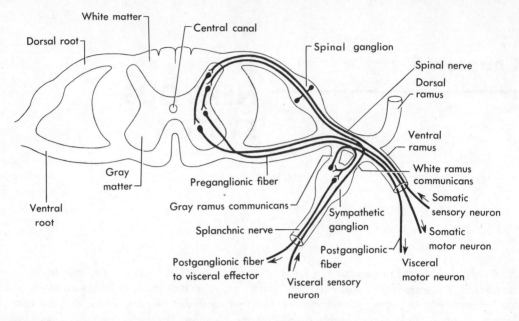

Figure 6-1 Diagram of a cross section through the spinal cord and a pair of spinal nerves to show the organization and functional components of the nervous system. Two internuncial neurons are shown in the gray matter between the sensory and motor neurons. (From Villee, Walker, and Barnes: General Zoology.)

white matter. The gray matter contains the cell bodies of neurons and unmyelinated fibers; the white, myelinated fiber tracts. Most of the gray matter is centrally located. In the cord, it forms a continuous column that has the appearance of an H, or a butterfly, in cross section. In the brainstem, the gray matter tends to break up into distinct nuclei, but the nuclei have a relationship to each other similar to the relationship between the parts of the gray matter in the cord. In higher vertebrates, some gray matter migrates to the surface parts of the brain where it forms a gray *cortex*.

The *peripheral nervous system* consists of all the neural structures—ganglia and nerves—lying outside the spinal cord and brain. *Spinal nerves* are segmentally arranged, and each connects to the central nervous system by a *dorsal* and a *ventral root* (Fig. 6-1). The dorsal root bears a *ganglion* containing the cell bodies of sensory neurons. More distally, the spinal nerve breaks up into branches, or rami, going to various parts of the body — a *dorsal ramus* to the epaxial region, a *ventral ramus* to the hypaxial region, and often one or more *communicating rami* with visceral connections. Although most of the *cranial nerves* may have had this pattern at one stage in their evolution, their segmentation and organization into dorsal and ventral roots are not so apparent in living vertebrates as those of the spinal nerves.

The *autonomic* portion of the peripheral nervous system is usually considered to consist of the motor neurons going to the visceral organs, glands, and smooth muscle generally. These leave the cord or brain in certain spinal and cranial nerves. Some of the fibers of the autonomic system remain in the spinal and cranial nerves, but some leave to travel in special branches to the organs in question. Thus, the autonomic system, while clearly definable functionally, is not completely separated from the rest of the peripheral nervous system morphologically.

A unique feature of the autonomic system is that there is a peripheral relay (Fig. 6-1). A *preganglionic neuron*, having its cell body in the gray matter of the central nervous system, leaves through the ventral root of a spinal or cranial nerve and goes to a peripheral ganglion. There it synapses with a *postganglionic fiber* that continues to the visceral organ. In contrast, only one neuron is involved in the innervation of the branchiomeric and somatic muscular units.

Another unique feature of the autonomic nervous system is its subdivision, in the higher vertebrates at least, into *sympathetic* and *parasympathetic* portions. A given visceral organ receives both types of fibers. One activates the organ; the other inhibits it. Sympathetic innervation increases the activity of the heart, slows down digestive processes, and has other effects that help the body to adjust to conditions of stress. Parasympathetic innervation has the opposite effect. The sympathetic fibers of the autonomic system leave through the thoracic and anterior lumbar spinal nerves; the parasympathetic, through certain cranial and sacral nerves. The peripheral relay of the sympathetic fibers is in a ganglion at some distance from the organ being supplied, so the postganglionic fiber is quite long. The parasympathetic relay, on the other hand, is in or very near the organ being supplied; thus its postganglionic fibers are relatively short. Glands in the skin receive only a sympathetic innervation. Many of the sympathetic ganglia lie against the back of the body cavity, lateral and ventral to the vertebral column. Those on each side of the body are interconnected by visceral fibers to form a chain known as the *sympathetic cord*.

FUNCTIONAL COMPONENTS AND THEIR INTERRELATIONS

A nerve may contain both *sensory (afferent)* and *motor (efferent) neurons.* These are mixed in the distal parts of a nerve, but they tend to segregate near and within the central nervous system. In a typical spinal nerve of a higher vertebrate, for example, the cell bodies of the motor neurons lie within the ventral portions of the gray matter of the cord, and their fibers leave through the ventral root of the spinal nerve (Fig. 6–1). The sensory neurons approach the cord through the dorsal root of the spinal nerve, and their cell bodies are located in the dorsal root ganglion. The sensory neurons ultimately enter the dorsal portions of the gray matter, but they may travel for some distance in the white matter of the cord before doing so.

Besides distinguishing between sensory and motor neurons, it is possible to distinguish between sensory neurons coming from somatic receptors and those coming from visceral receptors. The same is true for motor neurons. Thus, there are four major types of neurons within the peripheral nerves—*somatic sensory, visceral sensory, visceral motor,* and *somatic motor*. These four types of neurons also correlate with areas of the gray matter in that they begin, or end, in definite regions, or columns. Somatic sensory neurons end in the most dorsal part of the gray matter (Fig. 6–1), which constitutes a *somatic sensory column*; visceral sensory neurons, just ventral to them in a *visceral sensory column*. The cell bodies of the visceral motor neurons constitute a *visceral motor column* located in the lateral part of the gray matter; those of the somatic motor neurons constitute a *somatic motor column* located in the most ventral part of the gray matter. Certain of these four categories may be further subdivided. For example, a distinction can be made between the visceral motor fibers to the viscera (autonomic system) and those to the branchiomeric musculature. The former are *general visceral motor fibers* and the latter *special visceral motor fibers*.

As explained, somatic and visceral sensory neurons ultimately enter their respective columns of the gray matter. These primary sensory axons typically send branches ascending and descending along the spinal cord white matter. This arrangement enhances the formation of multiple synaptic connections responsible for both simple and complex reflexes, the latter always involving *internuncial neurons* between the sensory input and motor output. These internuncial neurons may pass through the gray matter to a motor neuron (a three-neuron reflex mechanism), or they may ascend in the white matter to the brain. Here they may form reflex connections or be relayed to still other parts of the brain. Internuncial neurons that ascend to the brain are called *afferent internuncial neurons*, and at some point during their ascent they usually cross (decussate) to the side of the central nervous system opposite to the one on which they originated. Impulses that originate in the brain descend through the white matter to the motor neurons on *efferent internuncial neurons*. These, too, usually decussate at some point during their descent.

DEVELOPMENT OF THE BRAIN

As might be expected, the brain is the most complex part of the nervous system. An understanding of the various regions of which it is formed is best gained from a consideration of its embryonic development. As shown in Figure 6–2, the brain arises as an enlargement of the anterior end of the neural tube. Very soon it becomes divided by certain constrictions into three regions: an anterior forebrain or *prosencephalon*; a middle midbrain or *mesencephalon*; and a posterior hindbrain or *rhombencephalon*.

The mesencephalon does not divide further, but the other two regions do. The prosencephalon divides into an anterior *telencephalon* and a posterior *diencephalon*; the rhombencephalon into an anterior *metencephalon* and a posterior *myelencephalon*. Each of these five regions further differentiates. The telencephalon gives rise to the cerebral hemispheres and olfactory bulbs; the diencephalon, to the thalamus, hypothalamus, epithalamus, and optic vesicles; the mesencephalon, to the rostral colliculi (optic lobes) and caudal colliculi which are located in its roof (tectum); the metencephalon, to the cerebellum and pons; the myelencephalon, to the medulla oblongata.

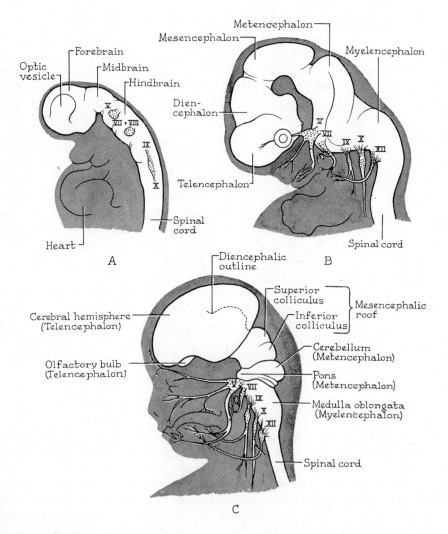

Figure 6–2 Diagrams of three stages in the development of the human brain to show the differentiation of the principal brain regions. (From Villee, Walker, and Barnes: General Zoology. Modified after Patten.)

EVOLUTION OF THE BRAIN

In the evolution from fish to mammals, numerous changes occur in the nervous system which are related in large measure to the increased activity and flexibility of response of mammals. The cerebrum of fish is primarily a center for olfactory integration, and few somatic sensory impulses reach it. The "master integrating" center of fish is the optic tectum, for sensory information from many parts of the body is projected here, and appropriate motor responses are initiated. A major change in the evolution of the mammal brain is the development of a *neopallium,* or neocortex, in the cerebral hemispheres to which sensory impulses are projected and where many motor impulses originate. Although a neopallium may have first appeared in reptiles (Fig. 6–3, *D*), it has not yet been definitely demonstrated in any living reptiles. It is, however, present in all mammals, although its extent in different groups varies considerably. As the neopallium develops, it causes a great expansion of the cerebrum. It pushes apart the paleopallium and archipallium, which originally were olfactory regions, as it comes to occupy the greater part of the cerebrum (Fig. 6–3). Enlargement of the neopallium also leads to a caudal growth of the cerebrum over the diencephalon and the mesencephalon. A similar caudal expansion occurs during embryonic development (Fig. 6–2). With the increasing dominance of the neopallium as the primary integrating structure, the tectum is left with a few optic and auditory reflex centers.

Another important change is the great enlargement of the cerebellum. This is correlated with the increased complexity of muscular movement, an increased projection of sensory data to this region, and an interconnection of cerebral hemispheres and cerebellum. To the original auricles (vestibular in nature) and body (locomotor) are added a pair of *hemispheres* related to the more delicate and precise musculoskeletal movements.

Other changes in the brain tend to be correlated with the increased importance of the neopallium and cerebellum. The thalamus enlarges as it becomes an important pathway and relay station between the cerebrum and other parts of the central nervous system. Other interconnections between the thalamus and cerebrum suggest that these two parts of the brain may form a complex interacting mechanism, so that the thalamus appears to be more than just a relay station. Important centers develop in the ventrolateral portions of the midbrain (red nucleus) and metencephalon *(pons)* for the interconnection of the cerebellum and cerebrum. More and larger fiber tracts, both afferent and efferent, evolve in the mesencephalon and hindbrain because the cerebrum, the cerebellum, and the brain, in general, exert more influence over the body than they did in lower vertebrates.

MENINGES

The mammalian brain and the stumps of the cranial nerves should be studied from isolated sheep brains. If this is not possible, the brain may be removed from your own specimen. In this case, postpone this study until the end of the course. To remove the brain, first make a sagittal section of the head, and then carefully loosen the halves of the brain and pull them out. Leave on the tough membrane covering the brain (dura mater) as you take it out. The cranial nerves will, of course, have to be cut, but leave stumps as long as possible. The foramina through which they leave the cranium are listed in Table 2–2 (p. 26), and certain of the nerves will be seen during later dissections.

The tough outer membrane that covers the brain is known as the *dura mater*. Actually, it represents the dura of lower vertebrates united with the periosteum lining the cranial cavity. Carefully remove the dura in order to see the other membranes. As you do so note that it sends one extension down between the cerebrum and cerebellum, and another between the two cerebral hemispheres. The former extension is call the *tentorium*; the latter, the *falx cerebri* (Fig. 7–7, p. 129). These membranes help

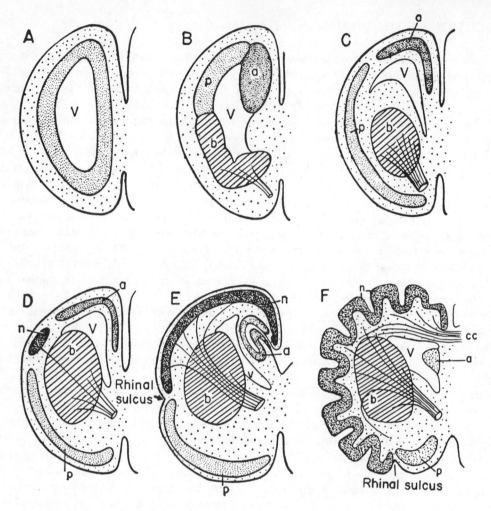

Figure 6–3 Diagrammatic cross sections of the left cerebral hemisphere showing important stages in its evolution. *A*, A primitive fish stage in which the hemisphere serves only as a central olfactory structure, and all the gray matter lines the ventricle, with very little if any differentiation of parts; *B*, a situation similar to that of today's living urodeles, probably a very generalized tetrapod condition, with differentiation of neural elements into three distinct groups; *C*, early amniote stage, in which pallial portions of gray matter migrate toward the surface but the corpus striatum remains internal and has increased fiber connections, other than those for olfaction; *D*, a mammal-like reptile in which the neopallium is beginning to appear; *E*, a primitive mammal, possibly a stage reached in the mid-Cretaceous, in which the neopallium becomes more important and pushes original pallial areas apart; *F*, a modern placental mammal in which the neopallium is extremely developed and connected with the neopallium of the opposite side by way of the corpus callosum. Although drawn to the same size, there is a considerable increase in size between stages *D* and *F*. Abbreviations: *a*, archipallium; *b,* corpus striatum; *cc,* corpus callosum; *n,* neopallium; *p,* paleopallium; *V,* lateral ventricle. Major fiber tracts are shown by groups of lines. The lines passing through the corpus striatum represent the internal capsule. (From Romer: The Vertebrate Body. *F*, redrawn and slightly modified.)

to stabilize the brain in the cranial cavity and to protect it from distortion during sudden rotational movements of the head. The pia-arachnoid of lower tetrapods has separated into two layers in mammals, but they are hard to distinguish grossly. The *pia mater* is the vascular layer that closely invests the surface of the brain. The *arachnoid membrane* lies between the pia and dura and is most easily distinguished from the pia in the region overlying the grooves on the brain surface, for the arachnoid does not dip into them, whereas the pia does. In life the arachnoid membrane lies against the dura, but there is a definitive *subarachnoid space* between the arachnoid and pia which is crisscrossed by weblike strands of connective tissue—a feature that gives the name to the arachnoid membrane. *Cerebrospinal fluid* circulates in the subarachnoid space where it forms a liquid cushion around the central nervous system.

The liquid cushion around the central nervous system is very important, for it helps to protect and support the exceedingly soft and delicate nervous tissue. Cerebrospinal fluid gives the brain a great deal of buoyancy. It has been calculated, for example, that a human brain weighing 1500 g. outside the body has an effective weight of only 50 g. in situ. The cerebrospinal fluid also helps to provide the brain with carefully selected nutrients and other substances, for fewer substances leave the capillaries within the central nervous system than leave the capillaries elsewhere in the body.

EXTERNAL FEATURES OF THE BRAIN AND THE CRANIAL NERVES

(A) TELENCEPHALON

The paired cerebral hemispheres and the cerebellum are so large that little else is at first apparent in a dorsal view. Notice that the *cerebral hemispheres* are separated from each other by a deep longitudinal furrow which is known as the *longitudinal cerebral fissure*. Spread the dorsal parts of the cerebral hemispheres apart and note the thick transverse band of fibers that connect them. This is the *corpus callosum*, a neopallial commissure found only in placental mammals. The surface of each hemisphere is thrown into many irregular folds (the *gyri*), which are separated from each other by grooves (the *sulci*). A pair of *olfactory bulbs* project from the anteroventral portion of the cerebrum (Fig. 6–4). They can be seen best in a ventral view. The olfactory bulbs lie over the cribriform plate of the ethmoid bone and receive the groups of olfactory neurons from the nose. These neuron groups, which will have been pulled off the bulbs, constitute the first cranial, or *olfactory nerve* (I). A summary of the distribution and functional components of all the cranial nerves is given in Table 6–1. At most, only their stumps can be seen during this dissection. A whitish band, the *lateral olfactory tract*, extends at an angle caudally and laterally from each bulb. The ventral portion of the cerebrum, to which each olfactory tract leads, is known as the *piriform lobe*. It is separated laterally from the rest of the cerebrum by the *rhinal sulcus*. Our cerebral hemispheres are relatively much larger and more complexly folded than in the sheep, and the olfactory bulbs are smaller (Fig. 6–5), but other major features of the telencephalon are very similar.

The piriform lobe represents the *paleopallium*, and it is homologous to the olfactory lobes of lower vertebrates. The remaining olfactory portion of the primitive cerebrum, *archipallium*, has been

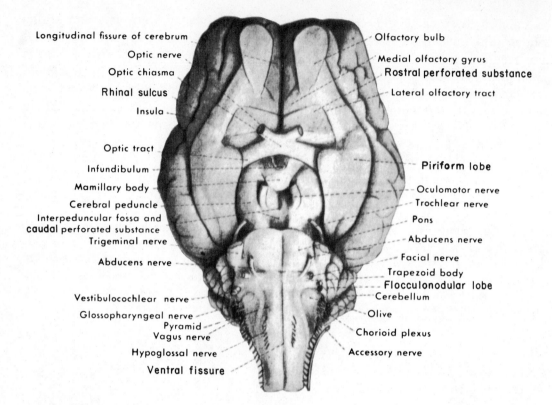

Longitudinal fissure of cerebrum
Optic nerve
Optic chiasma
Rhinal sulcus
Insula

Optic tract
Infundibulum
Mamillary body
Cerebral peduncle
Interpeduncular fossa and caudal perforated substance
Trigeminal nerve
Abducens nerve

Vestibulocochlear nerve
Glossopharyngeal nerve
Pyramid
Vagus nerve
Hypoglossal nerve
Ventral fissure

Olfactory bulb
Medial olfactory gyrus
Rostral perforated substance
Lateral olfactory tract

Piriform lobe

Oculomotor nerve
Trochlear nerve
Pons
Abducens nerve
Facial nerve
Trapezoid body
Flocculonodular lobe
Cerebellum
Olive
Chorioid plexus
Accessory nerve

Figure 6–4 Ventral view of the sheep brain. (From Ranson and Clark: Anatomy of the Nervous System. W.B. Saunders Company.)

pushed internally and does not show on the surface. Thus, all of the superficial part of the cerebrum that lies lateral and dorsal to the rhinal sulcus is *neopallium*—the major integrating region of the brain.

(B) DIENCEPHALON

The telencephalon has enlarged to such an extent that it has grown back over and covers the diencephalon and much of the mesencephalon. To see the dorsal portion of the diencephalon *(epithalamus),* it is necessary to spread the cerebral hemispheres apart, cutting the corpus callosum. Much of the roof of the diencephalon consists of a thin membrane from which a vascular *chorioid plexus* is suspended. This is one of several such plexuses which protrude into cavities within the brain. It is they that secrete the cerebrospinal fluid, part of which escapes to circulate in the subarachnoid space. Remove the chorioid plexus. The longitudinal slit that is then exposed is the *third ventricle*. The knoblike *pineal gland* lies posterior to the ventricle. It has been implicated in the regulation of sexual development, for secretions of this gland inhibit gonadal development in some species. Extracts of the organ (melatonin) also cause retraction of pigment in the chromatophores of lower vertebrates. Exposure to light decreases the activity of the gland. The narrow transverse band of tissue between the pineal gland and the ventrical is the *habenular commissure* and the tissue forming the posterolateral rim of the ventrical is the *habenula* (Fig. 6–7). The habenular commissure and the habenulae form the *habenular trigone.*

TABLE 6–1   *Distribution and Components of the Cranial Nerves**

Nerve	Distribution	Somatic Sensory Cutaneous	Somatic Sensory Proprioceptive	Somatic Sensory Special	Visceral Sensory General	Visceral Sensory Taste	Visceral Motor Autonomic	Visceral Motor Branchiomeric	Somatic Motor
Terminalis	Nasal epithelium	X					X		
I. Olfactory	Olfactory epithelium			X					
II. Optic	Retina			X					
III. Oculomotor	Ocular muscles: ventral oblique; ventral, dorsal, and medial rectus; levator palpebrae superioris		X						X
	Ciliary body of the eye						X		
IV. Trochlear	Ocular muscles: ventral oblique;		X						X
V. Trigeminal	Skin of scalp and face; teeth; tongue	X							
	Jaw muscles		X					X	
VI. Abducens	Ocular muscles: Lateral rectus, retractor bulbi		X						X
VII. Facial	Taste buds on rostral two-thirds of tongue					X			
	Facial muscles							X	
	Salivary and tear glands						X		
VIII. Vestibulocochlear	Inner ear			X					
IX. Glossopharyngeal	Taste buds on caudal third of tongue					X			
	Pharyngeal lining				X				
	Pharyngeal muscles							X	
	Salivary glands						X		
X. Vagus	Receptors in many internal organs: e.g., larynx, lungs, heart, aorta, stomach				X	X			
	Muscles of pharynx and larynx							X	
	Muscles of heart and gut; gastric glands						X		
XI. Accessory	Visceral arch muscles associated with the pectoral girdle: sternocleidomastoid, trapezius		X					X	
XII. Hypoglossal	Muscles of tongue		X						X

*The general distribution of mammalian cranial nerves is shown. An X indicates the presence of a particular type of neuron. The terminal nerve is a minor one, not visible grossly, and discovered only after the numbering system had become established.

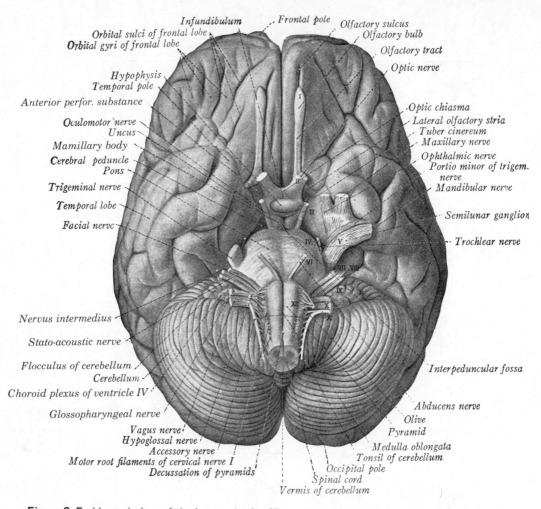

Figure 6–5 Ventral view of the human brain. (From Ranson and Clark: Anatomy of the Nervous System. W.B. Saunders Company. After Sobotta-McMurrich.)

Turn the brain over and examine the ventral surface of the diencephalon *(hypothalamus)*. The *optic nerves* (II) undergo a partial decussation at the anterior border of the hypothalamus, forming the prominent *optic chiasma.* The rest of the hypothalamus is the oval area lying posterior to the optic chiasma. The *hypophysis* (pituitary gland) may still be suspended by a narrow stalk, the *infundibulum*, from the hypothalamus. If so, remove it in order to get a clearer view of the region. The cavity in the infundibulum represents an extension of the third ventricle. That portion of the hypothalamus adjacent to the infundibulum is known as the *tuber cinereum*. A pair of rounded *mamillary bodies* forms the caudal end of the hypothalamus. These are more obviously paired in the human than in the sheep brain (Fig. 6–5).

In order to see the *thalamus* or lateral wall of the diencephalon, you must carefully pull one of the cerebral hemispheres forward and look beneath it. A better view will be had after the cerebrum has been dissected (p. 115), so you may care to postpone a study of the thalamus until then. An *optic tract* leads from the optic chiasma to terminate in an enlargement of the thalamus known as the *lateral geniculate body* (Fig. 6–6). The meninges will have to be removed to see the area clearly. The smaller enlargement posterior to the lateral geniculate body is the *medial geniculate body*, and that portion of the thalamus lying dorsal to the geniculate bodies is known as the *pulvinar*. These features are the same in a human brain.

The thalamus is an important relay station between the cerebral hemispheres and the rest of the brain. All sensory impulses, except for olfaction, pass through the dorsal portion of the thalamus on the way to the cerebrum. The medial geniculate body, for example, relays auditory impulses to the cerebrum. Most fibers in the optic tract terminate in the lateral geniculate body, from which impulses are relayed to the visual cortex. Some optic fibers go to the rostral colliculus, from which impulses go to the pulvinar and thence to the visual cortex. The ventral portion of the thalamus, which is not exposed on the surface, relays certain efferent impulses on the way back from the cerebral hemispheres. In addition, the thalamus may act as a subcortical center of integration, and its numerous interconnections with the cortex suggest that it plays a role in many cortical functions. The hypothalamus is an important integrating center for many autonomic and visceral functions, including sleep, body temperature, digestion, blood sugar level, water balance and sexual activity. The habenula is an important olfactory center. Impulses reach it from the piriform lobe of the cerebrum and go out to nuclei in the floor of the mesencephalon.

(C) MESENCEPHALON

The roof, or *tectum*, of the mesencephalon can be seen by spreading the cerebrum and cerebellum apart. Four prominent, round swellings (the *corpora quadrigemina*) characterize this region. The larger, rostral pair are the *rostal* (superior) *colliculi*; the smaller, caudal pair, the *caudal* (inferior) *colliculi*. Note that the *trochlear nerves* (IV) arise slightly posterior to the caudal colliculi (Fig. 6–6).

A pair of *cerebral peduncles* lies along the ventrolateral surface of the mesenceph-

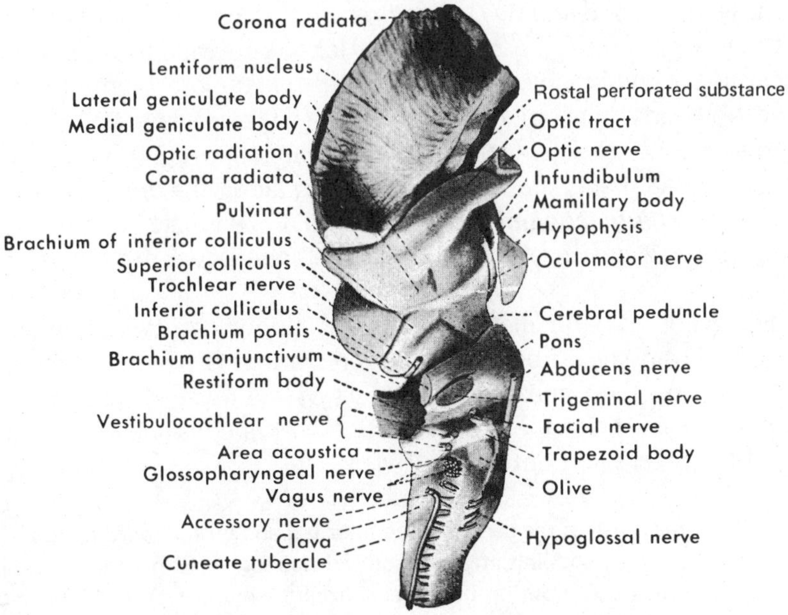

Figure 6–6 Lateral view of the sheep brainstem. (From Ranson and Clark: Anatomy of the Nervous System. W.B. Saunders Company.)

alon. Each emerges from beneath the optic tract and is as wide as the distance from the medial geniculate body to the hypothalamus. An *oculomotor nerve* (III) arises from the surface of each peduncle. The depression between the two peduncles is the *interpeduncular fossa*. If you strip the meninges from this region, you may be able to see small holes through which blood vessels enter the brain. This region constitutes the *caudal perforated substance*. A comparable *rostral perforated substance* lies anterior to the optic chiasma (Fig. 6–4). The human mesencephalon is the same.

The evolution of the neopallium has robbed the tectum of its original importance as the major integrating area, but optic and auditory fibers are still projected to the rostral and caudal colliculi, which remain significant visual and auditory reflex centers, respectively. The peduncles are large bundles of fibers that extend caudally from the cerebral hemispheres. Most efferent impulses from the cerebrum pass back through them.

(D) METENCEPHALON

The dorsal portion of the metencephalon forms the *cerebellum*. It will be noted that the surface area of the cerebellum is increased by numerous platelike folds *(folia)*, separated from each other by *sulci*. The median part of the cerebellum, which has the appearance of a segmented worm bent nearly in a circle, is called the *vermis;* the lateral parts are the *hemispheres*. The lobe of each hemisphere that lies ventral to the main part of the hemisphere, and lateral to the region where the cerebellum attaches to the rest of the brain, is known as the *flocculonodular lobe.* These lobes are homologous to the auricles of lower forms and receive vestibular impulses; most of the vermis is homologous to the body; and most of the hemispheres are new additions with cerebral connections.

The cerebellum is connected with other parts of the brain by three prominent fiber tracts, or peduncles (Figs. 6–6 and 6–7). The *brachium pontis,* or middle peduncle, lies medial to the anterior half of the flocculonodular lobe. You will have to dissect off this lobe on one side to see the brachium clearly. Note that the brachium pontis connects ventrally with a transverse band of fibers known as the *transverse fibers of the pons.* The tissue posterior and slightly medial to the brachium pontis constitutes the *restiform body,* or caudal peduncle. It also continues along the dorsolateral margin of the medulla. The *brachium conjunctivum,* or cranial peduncle, lies medial to the brachium pontis and can be seen by looking in the area between the cerebellum and caudal colliculi.

The ventral portion of the metencephalon is known as the *pons.* Grossly, it includes the region of transverse fibers; it is bordered anteriorly by the interpeduncular fossa, and posteriorly by the trapezoid body (part of the myelencephalon). The large *trigeminal nerve* (V) arises from the lateral portion of the pons and extends anteriorly across the base of the brachium pontis (Fig. 6–4).

The cerebellum in all mammals is the center for equilibrium and motor coordination. It monitors the motor activity of the body and initiates corrective impulses. In this connection, it receives vestibular fibers from the inner ear, proprioceptive fibers from the muscles of the body, and a variety of impulses from the cerebrum. Fibers from the cerebrum are relayed in the pons, decussate in the transverse fibers of the pons, and ascend to the cerebellum through the brachium pontis. Proprioceptive fibers come in through the restiform bodies. Vestibular fibers also enter in this region.

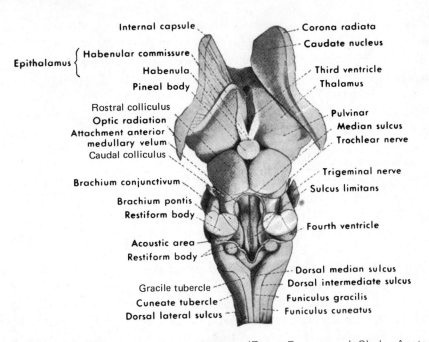

Internal capsule — — Corona radiata

— Caudate nucleus

Epithalamus { Habenular commissure

Habenula — Third ventricle

Pineal body — Thalamus

Rostral colliculus

Optic radiation — Pulvinar

Attachment anterior — Median sulcus
medullary velum — Trochlear nerve

Caudal colliculus

Brachium conjunctivum — Trigeminal nerve

— Sulcus limitans

Brachium pontis

Restiform body

— Fourth ventricle

Acoustic area

Restiform body

— Dorsal median sulcus

Gracile tubercle — Dorsal intermediate sulcus

Cuneate tubercle — Funiculus gracilis

Dorsal lateral sulcus — Funiculus cuneatus

Figure 6–7 Dorsal view of the sheep brainstem. (From Ranson and Clark: Anatomy of the Nervous System. W.B. Saunders Company.)

Most efferent fibers from the cerebellum pass out through the brachium conjunctivum to the ventral portion of the midbrain and metencephalon. Here they are relayed to the thalamus and to the motor columns.

(E) MYELENCEPHALON

All the rest of the brain belongs to the myelencephalon and forms the *medulla oblongata.* Its basic structure is the same in the sheep and human brain. In order to see the parts of the medulla clearly, the meninges must be stripped off on at least one side, but the remaining cranial nerves should be identified before this is done. At the border between the pons and the medulla oblongata are the *abducens* (VI) and *facial* (VII) *nerves;* the abducens lies more medial and extends anteriorly across the pons, and the facial passes out laterally with the eighth nerve. The stump of the *vestibulocochlear nerve* (VIII) lies dorsal to the seventh nerve and ventral to the posterior end of the flocculonodular lobe. The *glossopharyngeal* (IX) and *vagus* (X) *nerves* are represented by a number of fine rootlets posterior to, but in line with, the eighth nerve. Since one cannot trace these rootlets into the peripheral parts of the nerves, it is impossible to say more than that the anterior rootlets belong to the glossopharyngeal and the posterior ones to the vagus. The *accessory nerve* (XI) is the large longitudinal nerve posterior to the vagus. It arises by a number of fine rootlets from the caudal end of the medulla and the cranial end of the spinal cord. The rostral end of the nerve is the end that leads out of the skull to the muscles that the nerve supplies. If the nerve also has a cut caudal end on your specimen, all its origin is not intact. The *hypoglossal nerve*

(XII) is represented by the rootlets on the posteroventral portion of the medulla. If any of the cranial nerves cannot be found, refer to Figure 6–4.

Now strip off the meninges on one side of the medulla, and the thin membrane with its *chorioid plexus* that forms much of the roof of the medulla. Pull the cerebellum forward and note the large *fourth ventricle* extending forward into the metencephalon. Only the caudal part of the roof of the ventricle contains the chorioid plexus; the rostral part of the roof is formed by a thin layer of fibers termed the *rostral medullary velum,* which can be seen by pushing the cerebellum caudally. The trochlear nerve decussates in the velum.

Note the enlargement on the dorsal rim of the medulla just caudal to the point at which the restiform body turns into the cerebellum, and dorsal to the vestibulocochlear nerve. Observe that it extends medially to an oval-shaped enlargement in the ventrolateral part of the floor of the fourth ventricle. This enlargement constitutes the *area acustica.*

Examine the dorsal surface of the caudal end of the medulla. (The medulla extends as far caudad as the *first spinal nerve.*) The prominent *dorsal sulcus* of the cord continues onto the medulla nearly to the fourth ventricle (Fig. 6–7). A less distinct *dorsal intermediate sulcus* lies about one-third centimeter lateral to the preceding, and a *dorsal lateral sulcus* slightly lateral to the dorsal intermediate. These grooves outline two longitudinal fiber tracts, a dorsal *fasciculus gracilis* and a more lateral *fasciculus cuneatus.* The rostral end of the former tract expands slightly to form a structure called the *gracile tubercle.* The latter has a comparable enlargement known as the *cuneate tubercle.* These enlargements are more evident in a human than in the sheep brain. As already stated, the restiform body forms the dorsolateral rim of the medulla rostral to these tubercles. It then turns dorsally to enter the cerebellum.

Turn the brain over, and examine the ventral surface of the medulla. The narrow, transverse band of fibers immediately caudal to the pons is the *trapezoid body.* Notice that its fibers can be followed dorsolaterally into the area acustica. The midventral groove is the *ventral fissure.* The longitudinal bands of tissue on either side of it that are approximately two-thirds centimeter wide are known as the *pyramids.* Note that some of the pyramidal fibers lie superficial to the trapezoid body. An area known as the *olive* lies lateral to the pyramids, caudal to the trapezoid body, and ventral to the glossopharyngeal and vagus nerves.

The medulla is a transitional region between the anterior parts of the brain and the spinal cord. Most of its gray matter represents the forward continuation of the columns of the cord and the break-up of these into nuclei. Numerous reflex activities occur between these nuclei, and many important visceral activities are controlled here: respiratory movements, salivation, swallowing, rate of heart beat, and blood pressure. Most of the nuclei cannot be seen grossly, but some form bulges on the surface. The area acustica is a region in which auditory fibers from the ear undergo various relays. Some of the relays extend auditory impulses through the trapezoid body. The olive represents another nucleus, but its function is uncertain. It has connections with the cerebellum via the restiform bodies and may be related to muscular coordination. Much of the white matter of the medulla represents fiber tracts passing through the region. The pyramids, for example, are a caudal continuation of the fibers that formed the cerebral peduncles. The fasciculi gracilis and cuneatus are largely composed of ascending proprioceptive fibers. These relay in nuclei in the gracile and cuneate tubercles before going to the thalamus (for relay to the cerebrum), or to the cerebellum via the restiform bodies.

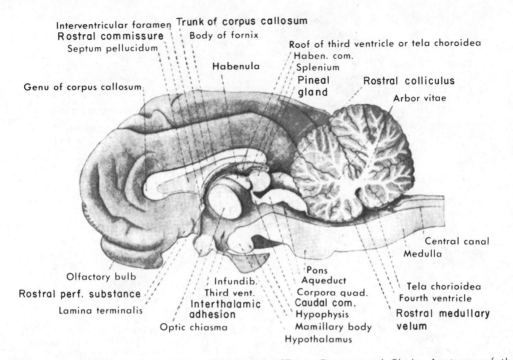

Figure 6-8 Sagittal section of the sheep brain. (From Ranson and Clark: Anatomy of the Nervous System. W.B. Saunders Company.)

SAGITTAL SECTION OF THE BRAIN

Cut the brain in half as close to the sagittal plane as possible. If you deviate from the plane, take the larger half and dissect away enough tissue to be able to see the median cavities of the brain clearly (Figs. 6–8 to 6–10). Many of the features just described can also be seen in this view. Note in particular the way in which the cerebral hemispheres extend back over the diencephalon and mesencephalon. The *corpus callosum* also shows particularly well. Its expanded anterior end is known as the *genu;* its expanded posterior end, as the *splenium;* and the thinner region between, as the *trunk of the corpus callosum.* A thin, vertical septum of tissue, *septum pellucidum,* lies ventral to the anterior part of the corpus callosum. It consists of two thin plates of gray matter. The lateral ventrical lies lateral to the septum and may be seen by breaking it. A band of fibers called the fornix lies posterior to the septum pellucidum. The *body of the fornix* begins near the splenium, and then the band curves forward and ventrally as the *column of the fornix.* It passes out of the plane of the section posterior to a small round bundle of fibers, which is a cross section of the *rostral* (anterior) *commissure,* an olfactory decussation. The fornix is a pathway that connects the cerebrum with the hypothalamus and will be understood better after the dissection of the cerebrum. The thin ridge of tissue extending ventrally from the rostral commissure to the optic chiasma is the *lamina terminalis*—a landmark representing the anterior end at the embryonic neural tube. The cerebral hemispheres are evaginations that extend laterally and then anteriorly from the portion of the brain lying just posterior to the lamina.

The third ventricle and diencephalon lie posterior to the column of the fornix, rostral commissure, and lamina terminalis. Note that the *third ventricle* is very narrow,

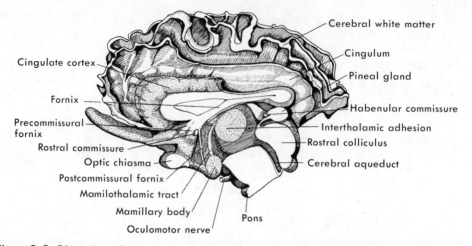

Cerebral white matter

Cingulum

Pineal gland

Cingulate cortex

Habenular commissure

Fornix

Interthalamic adhesion

Precommissural fornix

Rostral colliculus

Rostral commissure

Cerebral aqueduct

Optic chiasma

Postcommissural fornix

Mamilothalamic tract

Mamillary body

Oculomotor nerve

Pons

Figure 6–9 Dissection of a sagittal section of the cerebrum and diencephalon of a sheep brain.

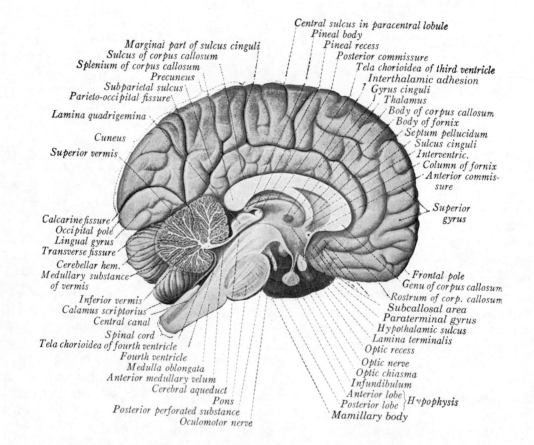

Central sulcus in paracentral lobule
Pineal body
Pineal recess
Posterior commissure
Tela chorioidea of third ventricle
Interthalamic adhesion
Gyrus cinguli
Thalamus
Body of corpus callosum
Body of fornix
Septum pellucidum
Sulcus cinguli
Interventric.
Column of fornix
Anterior commissure

Marginal part of sulcus cinguli
Sulcus of corpus callosum
Splenium of corpus callosum
Precuneus
Subparietal sulcus
Parieto-occipital fissure

Lamina quadrigemina

Cuneus

Superior vermis

Superior gyrus

Calcarine fissure
Occipital pole
Lingual gyrus
Transverse fissure
Cerebellar hem.
Medullary substance of vermis

Inferior vermis
Calamus scriptorius
Central canal

Spinal cord
Tela chorioidea of fourth ventricle
Fourth ventricle
Medulla oblongata
Anterior medullary velum
Cerebral aqueduct
Pons
Posterior perforated substance
Oculomotor nerve

Frontal pole
Genu of corpus callosum
Rostrum of corp. callosum
Subcallosal area
Paraterminal gyrus
Hypothalamic sulcus
Lamina terminalis
Optic recess
Optic nerve
Optic chiasma
Infundibulum
Anterior lobe
Posterior lobe } Hypophysis
Mamillary body

Figure 6–10 Sagittal section of the human brain. (From Ranson and Clark: Anatomy of the Nervous System. W.B. Saunders Co. After Sobotta-McMurrich.)

but that it has a considerable dorsal-ventral and anterior-posterior extent. It is lined by a shiny epithelial membrane (the *ependymal epithelium*), as are all the cavities within the neural tube. The thalamus lies lateral to the third ventricle, but a portion of it, the *interthalamic adhesion,* extends across the third ventricle and will appear as a dull, circular area not covered by the shiny ependyma. The *interventricular foramen,* or foramen of Monro, through which each lateral ventricle communicates with the third ventricle, lies in the depression anterior to the interthalamic adhesion. The hypothalamus lies ventral to the third ventricle and the epithalamus dorsal to it. Note again the *pineal gland,* the *habenular commissure* (which will show in cross section), and the *habenula.* These features show unusually well in the section. In addition, the epithalamus includes a *caudal* (posterior) *commissure.* This is the tissue ventral to the attachment of the pineal gland and anterior to the corpora quadrigemina.

After these features in the section have been seen, carefully dissect away tissue between the rostral commissure and the mamillary bodies. You will soon find a distinct band of fibers, the *postcommissural fornix,* which is continuous with the column of the fornix and leads to the mamillary bodies (Fig. 6–9). This is one of the main connections between the cerebrum and the hypothalamus. Dissecting just anterior to the rostral commissure, you will find fibers from the column of the fornix leading into the area of gray matter just under the septum pellucidum. This is the *septal region,* and these fibers in front of the rostral commissure constitute the *precommissural fornix.*

A narrow *cerebral aqueduct* (aqueduct of Sylvius) leads through the mesencephalon to the fourth ventricle of the hindbrain. The cerebellum lies above the fourth ventricle. Note that most of its *gray matter* is in the form of a gray *cortex* over the surface of the folia, while the *white matter* is centrally located. The white matter, which represents fiber tracts extending between the cerebellum and other parts of the brain, has the appearance of a tree and is called the tree of life *(arbor vitae).*

DISSECTION OF THE BRAIN
By DOUGLAS B. WEBSTER, Louisiana State University Medical Center

So far your study of the brain has involved primarily the identification of external landmarks, which, though a vital first step, gives but little feeling of the structural and functional continuity of parts. After you have mastered these landmarks, therefore, a dissection is necessary to gain appreciation of the brain's structural organization. The basic organization is the same in a sheep and human brain.

(A) CEREBRAL CORTEX

Take one half of your sheep brain, preferably the half on which you have demonstrated the precommissural and postcommissural fornix. With a pair of fine forceps carefully remove the pia and arachnoid layers from the cerebral hemisphere and cerebellum. Notice that the pia mater and its blood vessels dip deep into each sulcus of the neopallium (the convoluted part of the cerebral cortex). To realize the extent of the *gray matter* of the neopallium compared to the underlying *white matter* (i.e., fibers leading to, from, and within the neopallium), either cut off a one centimeter slice of the cerebrum in the frontal place or, if time permits, dissect off the gray matter. The dissection is best accomplished by scraping gently with a blunt instrument such as the handle of your scalpel or forceps or an orangewood manicure stick. Scrape deeply into each sulcus to remove all the gray matter, leaving only white. Do this for all the neopallium with the exception of the *insular cortex,* which will be removed later. The insular cortex is that part of the neopallium just above the rhinal sulcus at the level where the lateral olfactory tract meets the piriform lobe (Fig. 6–4).

When all this gray matter has been removed, notice the shape of the white matter deep to it. Especially notice how the fibers extend upward to form a huge core in each gyrus of the cerebrum. Also notice the large mass of tissue you have removed; this gives you a better idea of just how much of the brain is neopallium.

The white matter you are now looking at is a tangle of heterogeneous fibers. Some are short *arcuate fibers* connecting one gyrus with an adjoining gyrus. Others are long *association fibers* connecting one gyrus with cells many gyri away. Still others are *commissural fibers,* passing through the *corpus callosum* to connect right and left hemispheres. Some are *projection fibers: thalamocortical fibers,* ascending from the thalamus to the cerebrum; and *corticobulbar* and *corticospinal fibers,* leading from cells in the neopallium down to other brain or spinal cord structures. These many fibers are not organized into easily dissected groups; only in histological sections, and especially through the use of degeneration studies, can they be definitely traced and identified.

(B) CORPUS STRIATUM

The deeper part of the cerebral hemispheres, especially anteriorly, has other groups of gray matter interspersed with fibers, and hence is known as the *corpus striatum.* The gray matter can be divided grossly into a lateral portion (the lentiform nucleus) and a

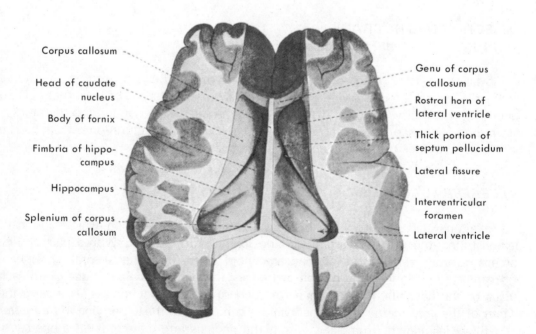

Corpus callosum

Head of caudate nucleus

Body of fornix

Fimbria of hippocampus

Hippocampus

Splenium of corpus callosum

Genu of corpus callosum

Rostral horn of lateral ventricle

Thick portion of septum pellucidum

Lateral fissure

Interventricular foramen

Lateral ventricle

Figure 6–11 Dorsal view of a dissection of sheep cerebrum. (From Ranson and Clark: Anatomy of the Nervous System. W.B. Saunders Company.)

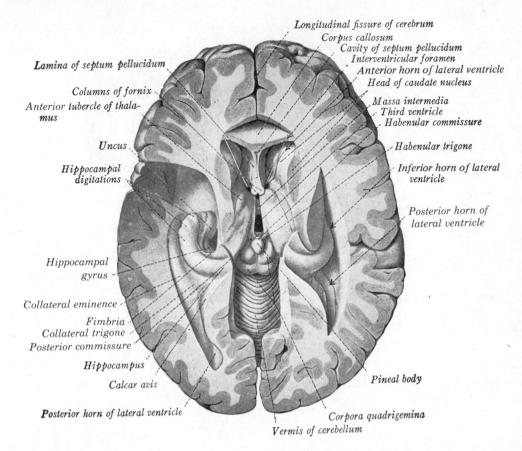

Longitudinal fissure of cerebrum
Corpus callosum
Cavity of septum pellucidum
Interventricular foramen
Anterior horn of lateral ventricle
Head of caudate nucleus

Lamina of septum pellucidum

Columns of fornix
Anterior tubercle of thalamus

Massa intermedia
Third ventricle
Habenular commissure

Uncus

Habenular trigone

Hippocampal digitations

Inferior horn of lateral ventricle

Posterior horn of lateral ventricle

Hippocampal gyrus

Collateral eminence
Fimbria
Collateral trigone
Posterior commissure

Hippocampus
Calcar avis

Pineal body

Posterior horn of lateral ventricle

Corpora quadrigemina
Vermis of cerebellum

Figure 6–12 Dorsal view of a dissection of the human brain. (From Ranson and Clark: Anatomy of the Nervous System. W.B. Saunders Co. After Sobotta-McMurrich.)

medial portion (the caudate nucleus) in each hemisphere. Carefully scrape away the reserved gray matter of the insular cortex until you come to a very thin layer of white fibers—the *external capsule.* Break through this external capsule and you will find a large mass of gray material; this is the *lentiform nucleus.* After identifying the lentiform nucleus and determining its boundaries, scrape it away. You will find a very large mass of fibers running between the thalamus and the deep white matter below the neopallium. This is the *internal capsule,* which contains the axons leading from neurons of the thalamus up to the cerebral hemisphere as well as the axons leaving the neopallium to pass down to other brain and spinal cord structures (Fig. 6–13).

The medial nucleus of the corpus striatum is best approached from the medial side of your brain half. Again locate the *septum pellucidum.* Cut it open if this has not already been done and look into the *lateral ventricle.* In doing this, be careful not to injure the part of the brain between the septum and the olfactory bulb. Note the very large head of the *caudate nucleus* (the medial portion of the corpus striatum), forming the ventrolateral border of the anterior horn of the lateral ventricle (Figs. 6–11 and 6-12). After identifying the large caudate nucleus, carefully scrape away its surface and see fibers of the internal capsule from the medial side. The caudate nucleus and the lentiform are penetrated and separated by the internal capsule, which connects neopallium with the rest of the brain. After passing between the nuclei, the fibers

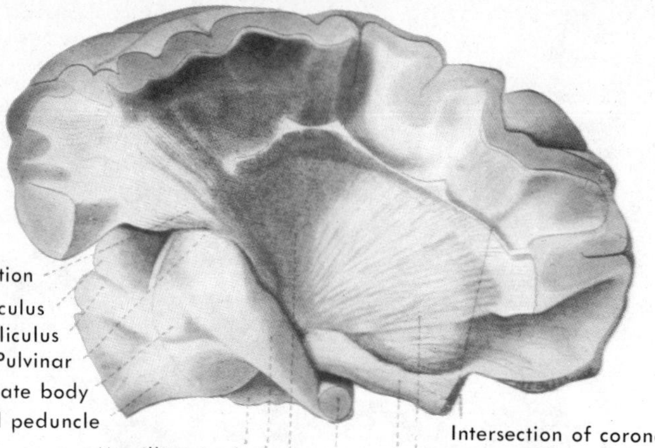

Figure 6–13 and labels:

Caudal thalamic radiation
Rostral colliculus
Caudal colliculus
Pulvinar
Medial geniculate body
Cerebral peduncle
Mamillary body
Optic tract
Caudal limb of internal capsule
Optic nerve
Intersection of corona radiata
and radiation of corpus callosum
Rostral limb of internal capsule
Rostral perforated substance

Figure 6–13 Lateral view of a dissection of the sheep brain to show the internal capsule. The lateral surface of the cerebrum and the lentiform nucleus have been removed. (From Ranson and Clark: Anatomy of the Nervous System. W.B. Saunders Company.)

of the internal capsule fan out to all parts of the neopallium in tracts called the *corona radiata* (Fig. 6–13).

(C) RHINENCEPHALON AND PAPEZ CIRCUIT

Now turn your attention to the rhinencephalon, or olfactory portions of the brain. During the evolution of the mammalian neopallium, outgrowths from the primitive olfactory system eventually developed into the bulk of the cerebral hemispheres. In the process the olfactory portions themselves were displaced (Fig. 6–3) and now follow circuitous routes. The olfactory nerves terminate in the olfactory bulbs, and each olfactory bulb, in turn, connects with more caudal parts of the cerebrum by means of three olfactory tracts or striae — the lateral, intermediate, and medial olfactory tracts.

The *lateral olfactory tract,* which you have already seen (Fig. 6–4), is the largest and most easily demonstrated. Careful scraping of the superficial gray matter will enable you to follow this tract into the *piriform lobe* and a small part of the adjoining neopallium. The smaller and shorter *intermediate olfactory tract,* which lies on the medial side of the preceding one, may be traced into the *rostral perforated substance,* where it terminates primarily in a nuclear concentration called the *olfactory tubercle.* Although in the sheep the olfactory tubercle is not grossly distinguishable from the rest of the anterior perforated substance, in many mammals its large surface bulges from the rostral perforated substance. By scraping away gray matter from the rhinencephalon near the midline and posterior to the olfactory bulb you will uncover the *medial olfactory tract.* Part can be traced to the *septal area* just under the genu of the corpus callosum; part, into the *rostral commissure.* The olfactory bulb also connects with the mamillary bodies, the tegmentum (or floor) of the midbrain, and the hippocampus, but these connections cannot be seen in gross dissection.

Prominent in the rhinencephalon and demonstrable by gross dissection is the *limbic lobe,* an important part of which is known as the *Papez circuit* (rhymes with "tapes"). In recent years the structures of the Papez circuit have come under a great deal of experimental scrutiny and they are now thought to be important in such diverse functions as emotional behavior and short-term memory.

Identify the *fornix* once again in the sagittal section, and follow its postcommissural portion to the *mamillary body.* Now carefully scrape away thalamic tissue just above the mamillary body caudal to the postcommissural fornix to reveal a tract running anterodorsally toward the anterior part of the thalamus. This is the *mamillothalamic tract.* Trace it up to the anterior part of the thalamus, where it ends in the *anterior thalamic nuclei;* it runs parallel to the postcommissural fornix (Fig. 6–9).

The anterior thalamic nuclei send fibers by way of the internal capsule and corona radiata into the cerebral cortex, and particularly into that part of the cerebral cortex which lies just above the corpus callosum. This cortical area, which can be seen on the undissected brain half, is called the *cingulate cortex.* It is possible to dissect this portion of the internal capsule, but since that would be destructive to other structures you will be seeing, this dissection should be saved for a later time. However, do identify the cingulate region where you have already removed cortex and notice there a distinct long association tract, called the *cingulum,* running anteroposteriorly, which is composed of fibers from cells of the cingulate cortex. Follow it caudally, around the back end of the corpus callosum, and then see it dip deep into the hemisphere where it will connect with the hippocampus.

The *hippocampus* is thought by some to be the most ancient cortex of vertebrates and is thus given the name *archipallium.* It lies in the floor of the lateral ventricle (Figs. 6–11 and 6–12) behind the caudate nucleus and will be seen best by spreading open the lateral ventricle where you have already cut the septum pellucidum. (The irregular-shaped, dark-colored frill in the floor of the ventricle is one of the *choroid plexuses.* It has invaginated from the thin, ventral wall of the hemisphere.) The surface of the hippocampus is covered by a very thin layer of white matter, called the *alveus.* Deep to this thin tract is the gray matter of the hippocampus. The cingulum enters it posteriorly, along its entire posterior border. The hippocampus extends lateroventrally, deep to the white matter of the neopallium, toward and almost reaching the piriform lobe. You may have to cut through the dorsolateral part of the hemisphere to trace it. The anterior border of the hippocampus lies free; along it is a tract of white matter called the *fimbria of the hippocampus.* Follow the fimbria of the hippocampus anteromedially and notice that it is continuous with and forms the body of the fornix.

You have now dissected the Papez circuit, going from the fornix by way of postcommissural fornix to the mamillary body, thence by way of the mamillothalamic tract to the anterior nucleus of the thalamus, thence by way of the internal capsule and corona radiata to the cingulate cortex, thence by way of cingulum back around to the hippocampus and, finally, by way of the fimbria, rejoining the fornix.

(D) CEREBELLUM

Cerebrum and cerebellum are also intimately connected to one another by circuit pathways, many of them demonstrable by dissection. Note how many of the lateral fibers of the internal capsule continue down into and form part of the *cerebral peduncle* and how these disappear into the pons. Within the pons, and you should verify this by careful scraping, is a great deal of gray material called the *pontine nuclei.* Many of the fibers coming down the cerebral peduncles terminate in synaptic connections with the cells of the pontine nuclei. Fibers from these cells of the pontine nuclei decussate in the pons and then enter the cerebellum (Fig. 6–6) by way of the *brachium pontis.*

Follow the brachium pontis into the cerebellum by scraping away part of the cerebellar hemisphere. These fibers continue up through the white matter of the cerebellum—the *arbor vitae*—and terminate in the cerebellar cortex (Fig. 6–8). Fibers leaving the cerebellar cortex follow the arbor vitae down as far as the base of the cerebellar peduncles and terminate there in large nuclear groups which may be demonstrated by dissecting shallowly into the lateral aspect of the brachium pontis.

The largest and most easily demonstrated of these cerebellar nuclei are the *dentate nuclei.* Their cells send fibers out of the cerebellum by way of the *brachium conjunctivum* (Fig. 6–6). These fibers cross the midline in the midbrain below the cerebral aqueduct, and many of them pass forward to terminate in the thalamus. In gross dissection the brachium conjunctivum can be seen leaving the cerebellum and entering the brainstem, but the mesencephalic and diencephalic portions can be visualized only in microscopic sections.

The fibers end in the lateral part of the dorsal thalamus, synapsing on thalamic cells which, in turn, send their fibers up through the internal capsule and corona radiata to the neopallium. As we see, therefore, there are distinct channels for information transfer and sharing between cerebrum and cerebellum by way of the cerebral peduncle, pontine nuclei, brachium pontis, cerebellar cortex, arbor vitae, cerebellar nuclei, brachium conjunctivum, thalamus, internal capsule, corona radiata, and neopallium.

(E) VISUAL SYSTEM

The cerebrum and cerebellum, the two main integrating centers of the brain, are concerned with information input and output—input being the information brought from each of the sense organs to the brain, and output the information sent out to skeletal muscles and other effectors. Since you have seen the structural connections between the cerebrum and cerebellum, it is now time to study the sensory channels leading to them and the motor pathways leaving them.

We have already examined the circuitous routes of the olfactory system, evolved as a by-product of the telencephalon's long phylogeny. Two other sensory systems—visual and auditory—lend themselves to gross dissections which will give you an idea of the structural apparatus for sending information up into the cerebral hemispheres.

The visual system is easily dissected, starting at the stumps of the two *optic nerves* where they enter the *optic chiasma.* Remember that the "optic nerve," strictly speaking, is not a true nerve but a brain tract containing axons of third-order neurons - from the ganglion cells of the retina. The optic nerve partially decussates in the optic chiasma: its fibers from the lateral half of the retina remain ipsilateral, while fibers from the medial half of the retina cross the midline.

Remove the posterolateral part of the cerebrum, including the hippocampus, but do so in such a way that you can place it back in position for further study. Remove the meninges from the lateral surface of the thalamus and mesencephalon. Follow the *optic tract* up the wall of the brainstem to the thalamic swelling known as the *lateral geniculate body* (Fig. 6–6). To see the gray matter of the lateral geniculate body, carefully lift the optic tract off the brainstem and peel it up and over the lateral geniculate body. You will notice many fibers penetrating the gray matter of the lateral geniculate body. Other, more superficial fibers pass over the lateral geniculate body and extend as the *brachium of the rostral colliculus* to terminate in the *rostral* (superior) *colliculus.* Still other fibers pass over the lateral geniculate body and enter the posterior thalamus in the region known as the *pulvinar,* whence they extend into the *pretectal region* between pulvinar and rostral colliculus. Careful dissection can reveal the *optic radiation,* the group of fibers that extend from the lateral geniculate body by way of the posterior part of the internal capsule and corona radiata to the *occipital neopallium.*

(F) AUDITORY SYSTEM

Now turn your attention to the auditory system and identify again the area where the eighth cranial nerve enters the brainstem. Here you will find, just behind the restiform body, the tubercle of gray matter known as the *area acustica,* composed of the *dorsal cochlear nucleus* and, deep to it, part of the *ventral cochlear nucleus* which also extends into the brainstem deep to the area acustica (Fig. 6–7). The auditory portion of the eighth cranial nerve terminates in these cochlear nuclei.

Proceeding ventrally from the area acustica, just behind the pons, identify again the *trapezoid body* as it passes around the ventral aspect of the brainstem and appears to join with its fellow on the other side just dorsal to the pyramids. Most, although by no means all, of the second-order auditory fibers leaving the cochlear nuclei cross the midline in the trapezoid body. Around the trapezoid body is a ventral nuclear group, unidentifiable by gross dissection, the *superior olivary complex,* one of the main termination points for second-order auditory neurons.

From here the auditory system moves forward as a major tract, the *lateral lemniscus,* which you will first see emerging between the anterior edge of the brachium pontis and the brachium conjunctivum and then running up into the *caudal* (inferior) *colliculus.* If you carefully peel off the brachium pontis it is possible to trace the lateral lemniscus all the way from the trapezoid body up into the inferior colliculus, where most or, more likely, all of its fibers terminate. From here, follow the *brachium of the caudal colliculus,* a prominent surface tract of fibers, as it crosses in an anteroventral direction down to a large thalamic prominence (Fig. 6–6). This surface swelling is the thalamic relay station of the auditory system, the *medial geniculate body;* the *auditory radiation* arises here, then passes into the internal capsule, and then by way of the corona radiata terminates in the *temporal neopallium.*

(G) PYRAMIDAL SYSTEM

The major motor system leaving the cerebral hemisphere is best picked up at the cerebral peduncle, which contains *corticospinal fibers* as well as the *corticopontine fibers* described before. Identify the cerebral peduncle once again and then, by scraping, follow it anterodorsally, up into the internal capsule and corona radiata. Its fibers begin in the motor cortex, perimotor cortical areas, and from areas of frontal and parietal cortex.

Now follow the cerebral peduncle caudally once again until it penetrates the pons. Further dissect the pons, demonstrating the gray as well as the white matter. Note that some of the fibers of the cerebral peduncle penetrate caudally directly through the pons and continue beyond it, forming the *pyramids.* Although much smaller than the cerebral peduncles, the pyramids contain the axons of the direct *corticospinal fibers* (Figs. 6–4 and 6–5). The pyramids continue caudally on both sides of the midventral line to the back part of the medulla oblongata, where most of their fibers decussate. Here you will see a slight bulge in the midventral surface of the brainstem. From this point on, the corticospinal fibers travel in different funiculi in different orders of mammals as they pass down the cord. However, they always terminate on the ventral horn cells of the spinal cord, and from these cell fibers go to the skeletal muscles, thus completing the output of motor information. The corticospinal fibers, with their cell bodies in the neopallium and their axons traveling all the way down to the ventral horn cells in the spinal cord, constitute the pyramidal system; this system, unique to mammals, is a major motor pathway for voluntary movements.

SPINAL CORD AND SPINAL NERVES

(A) THE CORD AND ROOTS OF THE SPINAL NERVES

The *spinal cord (medulla spinalis)* of the cat and human being is a subcylindrical cord of nervous tissue lying within the vertebral canal of the vertebral column. It is not uniform in diameter, for it bears *cervical* and *lumbosacral enlargements* from which nerves to the appendages arise, and the caudal end tapers as a fine *terminal filament* to end in the base of the tail. Only a section of the cord need be studied. To get at it, remove the epaxial muscles from your specimen so as to expose three or four inches of the vertebral column. This should be done in the thoracic region. With bone scissors, carefully cut across the pedicles of the vertebrae and remove the tops of the vertebral arches. The spinal cord will be seen lying in the vertebral canal. Continue chipping away bone and removing fat from around the cord, until you have satisfactorily exposed it along with several roots of the spinal nerves.

The cord and roots are covered by the tough *dura mater.* Note that the dura is not fused with the periosteum lining the vertebral canal as it is in the cranial cavity. Leave the dura on for the present and examine the roots of the spinal nerves. At each segmental interval, there is a pair of dorsal and ventral roots (Fig. 6–1). Trace a dorsal and ventral root laterally on one side. They pass into the intervertebral foramen before uniting to form a *spinal nerve.* Just before uniting, the dorsal root bears a small round enlargement—the *spinal ganglion* (dorsal root ganglion). If you trace the spinal nerve laterally you may see it divide into a *dorsal ramus* to the epaxial regions of the body and a *ventral ramus* to the hypaxial regions. The small *communicating rami* to the sympathetic ganglia and cord probably will not be seen.

Slit open the dura at one end of the exposed area. This opens the *subdural space.*

Observe that the roots of the spinal nerves do not have a simple attachment on the cord but unite by a spray of fine *rootlets.* Cut out a segment of the spinal cord and strip off the remaining meninges—*arachnoid* and *pia mater.* There is a deep ventral furrow on the cord known as the *ventral fissure;* a less distinct dorsal furrow, the *dorsal sulcus;* and a more prominent furrow slightly lateral to the middorsal line, the *dorsal lateral sulcus.* Note that the dorsal rootlets enter along the dorsal lateral sulcus. Recall that these same grooves extend onto the medulla oblongata.

Make a fresh cross section of the cord with some sharp instrument such as a razor blade. Examine the cut surface, and compare it with Figure 6–1. Also look at demonstration slides if possible. The tiny *central canal* can generally be seen grossly, and if you are fortunate you may be able to distinguish the butterfly-shaped central *gray matter* from the peripheral *white matter.* As explained in the introduction to this chapter, the gray matter consists of unmyelinated fibers and the cell bodies of motor and internuncial neurons; the white, of ascending and descending, myelinated fibers. That segment of the white matter that lies between the dorsal fissure and the dorsal lateral sulcus is called the *dorsal funiculus;* that segment between the dorsal lateral sulcus and the line of attachment of the ventral roots, the *lateral funiculus;* and that portion between the ventral roots and the ventral median fissure, the *ventral funiculus.*

(B) THE BRACHIAL PLEXUS

There are 38 spinal nerves in the cat (8 cervical, 13 thoracic, 7 lumbar, 3 sacral, 7 caudal). We have 31 spinal nerves (8 cervical, 12 thoracic, 5 lumbar, 5 sacral, 1 coccygeal or caudal). Their dorsal rami extend straight out into the epaxial region, but many of the ventral rami unite in a complex manner to form networks, or *plexuses,* before being distributed to the musculature and skin. This is especially true in the region of the appendages. In a typical mammal, the anterior cervical nerves form a *cervical plexus* supplying the neck region; the posterior cervical and anterior thoracic nerves form a *brachial plexus* supplying the pectoral appendage; and the lumbar, sacral, and anterior caudal nerves form a *lumbosacral plexus* supplying the pelvic appendage.

The brachial and lumbosacral plexuses of the cat may be dissected as examples of these networks. The *brachial plexus* lies medial to the shoulder and cranial to the first rib; it should be approached from the ventral surface. If it is still intact on the side on which the muscles were dissected, study it there; otherwise, cut through the pectoralis complex of muscles on the other side. The dissection of the plexus involves the meticulous picking away of fat and connective tissue from around the nerves and the accompanying blood vessels. If you find it necessary to cut any of the larger vessels, do so in such a way that you will be able to appose the cut surfaces when you study the circulatory system. Clean off the nerves from their point of emergence between the longus coli and scalenus muscles to the point at which they disappear into the shoulder muscles and brachium.

The brachial plexus is formed by the union of the ventral rami of the sixth to eighth cervical and first thoracic nerves in the cat. The fifth cervical also contributes to it in some mammals, including human beings. Considerable variation occurs in the details of the union of these nerves to form the plexus and in the origin of peripheral branches, but a common pattern is shown in Figure 6–14. The ventral rami of the nerves that enter the plexus are referred to as the *roots (radices)* of the plexus. It will

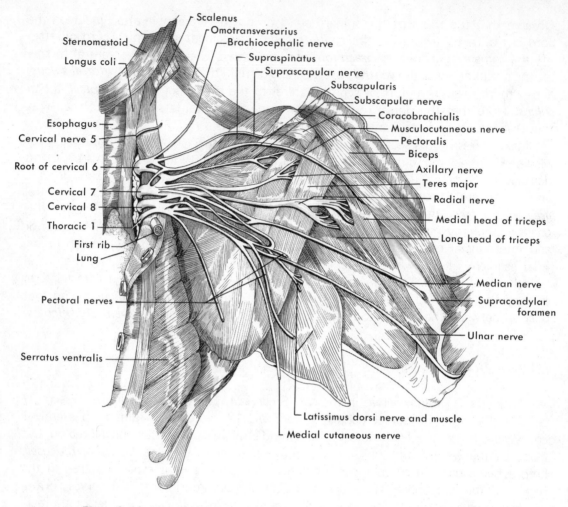

Figure 6–14   Ventral view of a dissection of the left brachial plexus of a cat.

be noted that each root tends to split into two *divisions* and that the divisions of different nerves unite to form *trunks* from which peripheral nerves arise. It will also be noted that the splitting of roots into divisions, and the union of divisions to form trunks, occurs in such a way that there tends to be an early segregation of the nerves supplying the dorsal appendicular muscles from those supplying the ventral. Many of the nerves are also cutaneous, being distributed to the skin, but only the major cutaneous branches will be described.

The most ventral nerves of the plexus are several *pectoral nerves.* They are small nerves which arise from the ventral divisions of the plexus, may or may not unite with each other, and pass to the pectoralis complex.

A *suprascapular nerve* leaves the front of the plexus, where it arises for the most part from the sixth cervical, and passes between the subscapular and supraspinatus muscles to supply the supraspinatus, infraspinatus, and some of the skin over the shoulder and brachium.

One or more *subscapular nerves* leave caudal to the suprascapular nerve and pass to the large subscapular muscle. They arise, for the most part, from the sixth and seventh cervicals.

A large *axillary nerve,* which lies caudal to the subscapular nerve, arises from the seventh cervical nerve. It passes between the teres major and the subscapularis to supply the teres minor deltoid complex and often the teres major.

Smaller nerves to the *latissimus dorsi* arise from the plexus near the origin of the axillary. They may also supply the teres major.

The large, deep nerve caudal to the axillary that is formed by the union of the dorsal divisions of the seventh and eighth cervicals and first thoracic is the *radial.* This is the largest nerve of the plexus. It passes through and between parts of the triceps, crosses the humerus to the lateral surface of the arm, and thence goes down the extensor portion of the forearm. It supplies the tensor fasciae antebrachii, triceps, and forearm extensors. A large branch of the nerve is cutaneous.

All the above nerves go to dorsal appendicular muscles, except the ventral thoracics and the suprascapular, which supply parts of the ventral musculature. The remaining nerves innervate the rest of the ventral appendicular muscles. It will be noted that all the nerves to the ventral muscles arise from the ventral divisions of the plexus. A small *musculocutaneous* springs from the sixth and seventh cervicals and enters the biceps. It may branch before reaching the biceps. It supplies the biceps, coracobrachialis, brachialis, and some of the skin over the forearm.

The ventral divisions of the seventh and eighth cervical and first thoracic combine to form two prominent nerves that run down the medial side of the brachium. The more cranial of these is the *median nerve,* the more caudal the *ulnar nerve.* They are distributed to the forearm flexors and skin of the hand. The median nerve passes through the supracondylar foramen of the humerus.

A small *medial cutaneous nerve,* which arises from the first thoracic, runs parallel with, and caudal to, the ulnar nerve. It supplies some of the skin over the forearm.

(C) THE LUMBOSACRAL PLEXUS

The *lumbosacral plexus,* which supplies the skin and muscles of the pelvis and hind leg, is located so deep within the abdominal and pelvic cavities that it cannot be dissected until the abdominal and pelvic viscera have been studied. If you are to dissect this plexus return to the following description after you have completed the urogenital system.

The pelvic symphysis will have been cut. Push the two hind legs dorsally thereby spreading open the pelvic canal, and push the abdominal and pelvic viscera to one side. Also cut the external iliac artery and vein shortly before they pass through the abdominal wall, and reflect them. Identify the psoas minor muscle (p. 67), and cut and reflect it. A longitudinal cleft can be found on the lateroventral part of the psoas major near the point where the deep circumflex iliac vessels cross it. Separate the psoas major along this cleft into superficial (ventral) and deep (dorsal) portions. Notice the nerves of the lumbosacral plexus emerging through this cleft, and dissect away the superficial portion of the psoas major as you trace them medially to the intervertebral foramina through which they leave the vertebral column.

The lumbosacral plexus of the cat if formed by the ventral rami of seven spinal nerves (the fourth lumbar to third sacral). As is the case with the brachial plexus, some

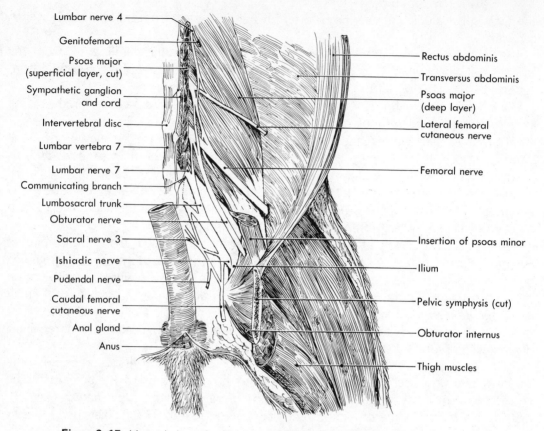

Lumbar nerve 4

Genitofemoral

Psoas major
(superficial layer, cut)

Sympathetic ganglion
and cord

Intervertebral disc

Lumbar vertebra 7

Lumbar nerve 7

Communicating branch

Lumbosacral trunk

Obturator nerve

Sacral nerve 3

Ishiadic nerve

Pudendal nerve

Caudal femoral
cutaneous nerve

Anal gland

Anus

Rectus abdominis

Transversus abdominis

Psoas major
(deep layer)

Lateral femoral
cutaneous nerve

Femoral nerve

Insertion of psoas minor

Ilium

Pelvic symphysis (cut)

Obturator internus

Thigh muscles

Figure 6–15   Ventral view of a dissection of the left lumbosacral plexus in a cat.

variation occurs in the way the seven roots of the plexus split to form divisions and the divisions unite to form the trunks from which the peripheral nerves arise. A common pattern is shown in Figure 6–15.

Lumbar nerve 4 splits soon after emerging from the intervertebral foramen into a *genitofemoral nerve* and a branch that passes caudad to join the divisions of the fifth lumbar nerve. The genitofemoral nerve continues caudad close to the external iliac vessels and passes through the body wall with the external pudendal vessels to supply the skin in the groin and on parts of the external genitalia. In a male small branches also go to the cremasteric muscle.

One division of *lumbar nerve 5* unites with the caudal branch of the fourth lumbar to form the *lateral femoral cutaneous nerve,* which extends laterally near the deep circumflex iliac vessels, perforates the body wall, and supplies the skin over the lateral surface of the hip and thigh. The other division of lumbar 5 continues caudad to unite with the divisions of the sixth lumbar nerve. Branches to the psoas muscles may arise from this division or from the lateral femoral cutaneous.

Lumbar 6 is a large nerve. Its largest division unites with a branch of the caudal division of the fifth lumbar to form the large *femoral nerve.* After perforating the body wall with the femoral artery and vein, the femoral nerve innervates the quadriceps femoris and certain other extensor thigh muscles. It also gives rise to a prominent cutaneous branch (the *saphenous nerve*) that supplies the skin on the medial side of the thigh and shin.

A second division of lumbar 6 receives another division of the fifth lumbar to form the *obturator nerve.* This nerve extends caudolaterally, perforates the obturator internus near the brim of the pelvis, and goes through the obturator foramen of the pelvic girdle to supply primarily the gracilis, obturator externus, and adductors of the thigh. These muscles belong to the ventral appendicular group.

A third division of lumbar 6 continues caudad to join *lumbar 7* and form the large *lumbosacral trunk.* As the lumbosacral trunk continues caudolaterally it soon receives a division from sacral nerve 1 and a bit farther distally a contribution from sacral 2. The trunk leaves the pelvic canal by passing between the ilium and sacrum, and breaks up into several branches. *Gluteal nerves* supply the gluteal and other laterodorsal hip muscles, and a large *ischiadic nerve* (sciatic nerve), which is the main continuation of the trunk, continues down the lateral surface of the thigh (Fig. 4–17, p. 64), innervating the biceps femoris, semimembranosus, semitendinosus, and other flexor muscles of the thigh. It bifurcates near the distal end of the thigh into *tibial* and *common peroneal* (fibular) *nerves.* These innervate the flexors and extensors of the shank, respectively. Cutaneous branches of the ischiadic, tibial, and common peroneal nerves help to supply adjacent skin.

Sacral nerves 1, 2, and 3 also interconnect with each other to form a network from which several small nerves arise, the most conspicuous being the pudendal and caudal femoral cutaneous nerves. The *pudendal nerve* contains motor fibers to striated muscles in the anal region and sensory fibers coming from the anal region and from the penis or clitoris. The *caudal femoral cutaneous nerve* helps supply the skin in the anal area and adjacent parts of the thigh.

The main peripheral branches of the human lumbosacral plexus are the same as those of the cat, but different spinal nerves contribute to them for we have five lumbar and five sacral nerves. The genitofemoral nerve is formed by branches from lumbar 1 and 2; the lateral femoral cutaneous from lumbar 2 and 3; the femoral and obturator nerves from lumbar 2, 3, and 4; the lumbosacral trunk from lumbars 4 and 5 and the first three sacral nerves; the posterior femoral cutaneous from sacral nerves 1, 2, and 3; and the pudendal nerve from sacrals 2, 3, and 4.

A portion of the *sympathetic cord* and *ganglia* can also be seen during this dissection lying on the ventral surface of the lumbar vertebrae (Fig. 6–15). Each ganglion receives a *communicating branch* from adjacent lumbar spinal nerves. Pelvic viscera receive their sympathetic innervation by minute branches from the pelvic extension of the sympathetic cord that follow the blood vessels to the organs. Parasympathetic innervation is from a *pelvic nerve* formed by very small branches from the sacral nerves. These nerves are seldom seen in dissections.

Chapter Seven

THE COELOM AND THE DIGESTIVE AND RESPIRATORY SYSTEMS

We now turn from the organ systems that support, move, and integrate the body's activities to a group that sustains metabolism. The digestive system brings in water, food, minerals, and other nutrients needed by the body; the respiratory system takes care of the essential gas exchanges; the circulatory system transports materials to and from the cells; and the excretory system eliminates the nitrogenous and many other waste products of cellular metabolism and helps to control the water balance of the body. These systems are functionally distinct, but it is convenient to study the digestive and respiratory systems together to some extent, for they are closely associated morphologically. The respiratory system develops embryonically as outgrowths from the digestive system.

THE COELOM AND ITS SUBDIVISIONS

The body cavity, or *coelom,* is an epithelium-lined space containing some liquid and surrounding the viscera. Its presence permits a certain amount of movement of the organs and their functional change in size and shape. Embryologically, the coelom develops as a pair of mesothelial-lined spaces within the lateral plate mesoderm on either side of the digestive tract (Fig. 7–1). At first there is a right and left coelom that converge above and below the digestive tract to form a *dorsal* and a *ventral* mesentery (Fig. 7–1, *B*); the dorsal portion, by a pair of folds (called *pericardiopleural membranes* in body, and the originally paired coeloms become continuous. It follows from this development that the viscera are technically outside the coelom, being separated from it by its thin mesothelial wall. However, the visceral organs are approached by cutting open the coelom, and it is therefore convenient to speak of them as being within the body cavity.

The coelom of fishes, amphibians, and early mammalian embryos is divided into two parts. The more cranial portion around the heart forms a *pericardial cavity,* and the rest a *pleuroperitoneal cavity.* The separation between them is a partition known as the *transverse septum.* The ventral portion of this septum develops embryologically through the expansion of the liver in the ventral mesentery (Fig. 7–1, *B*); the dorsal portion, by a pair of folds (called *pericardiopleural membranes* in mammalian embryology) that carry the ducts of Cuvier (common cardinal veins) down to the heart (Fig. 7–2). In fishes, the transverse septum is a vertical partition, but the caudad migration of the heart and pericardial cavity in most tetrapods causes the septum to assume a somewhat oblique position. The paired cranial parts of the pleuroperitoneal cavity thus come to lie dorsal to the pericardial cavity (Fig. 7–2). It is in these cranial recesses that the lungs lie. The separation of this pair of recesses as *pleural cavities* (when it eventually occurs in certain reptiles) is through the

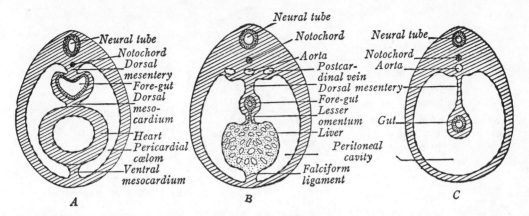

Figure 7–1 Diagrammatic cross section of a vertebrate embryo to show the relationship of the coelom and mesenteries to the visceral organ: *A*, through the level of the heart; *B*, through the liver; *C*, through the intestine. All mesenteries ventral to the digestive tract are parts of the ventral mesentery. (From Arey: Developmental Anatomy. After Prentiss.)

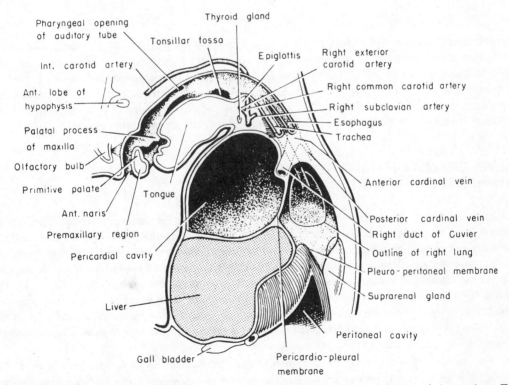

Figure 7–2 Sagittal section of a human embryo to show the subdivisions of the coelom. The pericardial cavity is already separated from the rest of the coelom, for the liver is expanding in the ventral portion of the transverse septum, and the dorsal part of the septum (pericardiopleural membrane) has developed to carry the ducts of Cuvier. The transverse septum has an oblique position, so the future pleural cavities (only the right one is shown) lie dorsal to the pericardial cavity. A pleuro-peritoneal membrane, which is shown developing, is largely responsible for the separation of each pleural cavity from the peritoneal cavity. (Redrawn from Hamilton, Boyd, and Mossman: Human Embryology. W. Heffer & Sons, Ltd.)

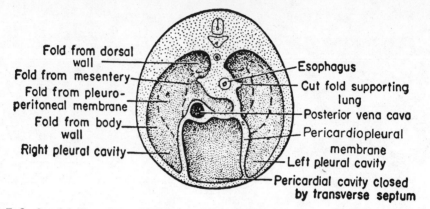

Figure 7–3 Cranial view of the diaphragm of the mammalian embryo to show the various folds of which it is composed. The heart and lungs have been removed to show the caudal portions of the two pleural cavities and the pericardial cavity (darker stippling). At this stage of development, the caudal surface of the pericardial cavity lies against the diaphragm, but it later separates from it in many mammals. The pleuropericardial membrane (also called pericardiopleural membrane) represents the dorsal part of the primitive transverse septum. The lighter stippling extending from dorsal to ventral is the mediastinum. (From Romer: The Vertebrate Body. After Broman and Goodrich.)

development of another pair of folds, the *pleuroperitoneal membranes* (Fig. 7–2), aided by subsidiary folds from the body wall and dorsal mesentery. The *diaphragm* of mammals represents these coelomic folds, the ventral part of the transverse septum, and somatic musculature of cervical origin that has invaded them (Fig. 7–3). In mammals, the pleural cavities subsequently extend ventrally lateral to the pericardial cavity and thus more or less surround the pericardial cavity. The part of the original pleuroperitoneal cavity caudal to the diaphragm now constitutes the *peritoneal cavity.*

DEVELOPMENT OF THE DIGESTIVE AND RESPIRATORY SYSTEMS

Muscle and connective tissue in the walls of the digestive and respiratory tracts develop from the visceral layer of the lateral plate mesoderm, but most of the epithelium lining these tracts, and the secretory cells of the glandular outgrowths, develop from the embryonic *archenteron,* and hence are lined with endoderm. However, variable amounts of the front and hind ends of the digestive tract are formed by ectodermal invaginations—the *stomodeum* and *proctodeum,* respectively (Fig. 7–4). The former forms the lining of the mouth or *oral cavity;* the latter contributes to the cloacal region. At first the ectodermal invaginations are separated from the archenteron by plates of tissue, but these eventually break down. It is then difficult to determine precisely where ectoderm ends and endoderm begins, and the precise limits of the oral cavity.

For purposes of description, the archenteron may be divided into a *foregut* and *hindgut.* The former differentiates into the pharynx, esophagus, and stomach; the latter differentiates into the intestinal region and much of the cloaca. In most mammals the cloaca is present only in embryos. It soon becomes divided—the dorsal part contributing to the rectum; the ventral, to the urogenital passages.

In all vertebrates, a series of *pharyngeal pouches* grow out from the side of the pharynx (Fig. 7–4). There are six of these in most fishes, fewer in tetrapods. The tissues between the pouches constitute the *branchial bars,* the first bar being cranial to the first pouch. The skeletal visceral arches, branchiomeric muscles, certain nerves, and the aortic arches grow into these bars. In fishes, the

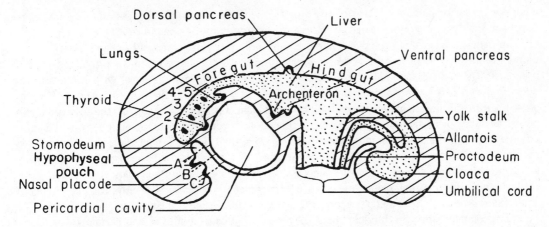

Figure 7–4 Diagrammatic sagittal section of a mammalian embryo to show the development of the digestive and respiratory systems. The points of entrance of the pharyngeal pouches are numbered. Lines *A*, *B*, and *C* indicate the comparative position of the mouth openings of an agnathous fish (*A*); a jawed fish not having internal nostrils (*B*); and fishes and tetrapods with internal nostrils (*C*).

endodermal pharyngeal pouches meet comparable ectodermal furrows, break through to the surface, and form the *gill slits* and *pouches* on whose walls the gills develop. The pharyngeal pouches do not normally break through in tetrapods (except for certain ones in larval amphibians), but are present embryonically nonetheless. The first gives rise to the *tympanic cavity* and *auditory tube;* the mammalian *tonsillar fossa,* containing the palatine tonsil, develops at the site of the second pouch; the other pouches give rise to certain glandular structures and then disappear. A *thymus* develops as epithelial thickenings of the ventral part of the third and fourth pouches in mammals. These primordia usually coalesce into a single one. The thymus has been implicated in the development of the body's capacity to produce antibodies. *Parathyroid glands* develop as epithelial thickenings from the dorsal part of the third and fourth pouches. Finally, *ultimobranchial bodies* develop in all vertebrates from the caudal face of the last pharyngeal pouch. They often join the thyroid primordium in mammals. Both the parathyroid glands and the ultimobranchial bodies produce hormones that regulate the level of calcium in the blood.

In addition to the lateral pharyngeal pouches, certain median evaginations arise from the floor of the pharynx. An endocrine *thyroid gland* grows out from the floor between the level of the first and second pouches and migrates caudad to the level of the larynx or trachea. The thyroid hormone increases the organism's metabolic rate.

The *lungs* of sarcopterygian fishes and tetrapods arise as a median bilobed evagination from the floor of the pharynx just caudal to the pharyngeal pouches. The early primordia of the lungs resemble a pair of pharyngeal pouches in the embryos of some amphibians, and it is possible that they evolved from a ventrally displaced pair of caudal pouches.

No other outgrowths arise from the digestive tract of most vertebrates until the level of the front of the hindgut. At this point one finds the liver and pancreas. The *liver* arises embryologically as a prominent ventral diverticulum (Fig. 7–4) which, as mentioned earlier, grows into the ventral mesentery caudal to the heart, and, by its expansion, forms the ventral part of the transverse septum. During subsequent development, the liver grows caudally in the ventral mesentery, and in the adult remains connected to the septum (or diaphragm) only by a mesentery known as the *coronary ligament.* Functionally, the liver is a very diverse organ. It secretes bile (a mixture of excretory products and the fat-emulsifying bile salts) and its cells come into intimate contact with blood that is brought to it from the stomach and intestinal region by the hepatic portal system. Many metabolic conversions occur here: excess absorbed sugars are stored, largely in the form of glycogen, or deficiencies are made up so the glucose content of the blood is kept at a constant level; amino acids are deaminated and their amino groups converted to urea; absorbed toxins may be removed; etc.

The pancreas arises embryologically from one or more intestinal outgrowths near the liver primordium. Frequently, there is both a dorsal evagination *(dorsal pancreas)* and a ventral one *(ventral pancreas)*. The latter is associated with the base of the liver anlage and, hence, with the future bile duct (Fig. 7–4). The ventral pancreas tends to work its way around the intestine, and to fuse and grow up into the dorsal mesentery with the dorsal pancreas. All the stalks of the primordia may persist as ducts, or certain ones may be lost in the adult. A ventral pancreatic duct can be recognized because it enters the intestine in common with the bile duct; a dorsal pancreatic duct, by its independent entrance on the opposite side of the intestine. Most pancreatic cells secrete digestive enzymes, which are discharged into the intestine and act on proteins, carbohydrates, fats, and nucleic acids. Little islands of endocrine tissue (the *islets of Langerhans*) are scattered among the exocrine cells; these islets produce insulin and glucagon, which are vital in carbohydrate metabolism.

More caudally along the hindgut there is, in the embryos of amniotes and certain fishes, a *yolk stalk* connecting with the *yolk sac* (Fig. 7–4). The yolk stalk and sac become relatively smaller as the embryo grows, and are lost by the adult stage.

A final major outgrowth is the *urinary bladder,* present in most tetrapods. It develops from the embryonic cloaca near the caudal end of the hindgut. In the embryos of amniotes, this structure expands considerably and extends beyond the limits of the embryo as the *allantois*—an extra-embryonic membrane which serves for excretion and respiration in the embryos of reptiles and birds, and for vascularizing the fetal portion of the placenta in eutherian mammals.

EVOLUTION OF THE DIGESTIVE AND
RESPIRATORY SYSTEMS

Evolutionary changes in the digestive and respiratory system are correlated first with the shift from water to land, which occurred between fishes and amphibians, and second with the evolution of a high and relatively constant rate of metabolism which enables mammals to be active, warm-blooded animals (endothermic). The latter change occurred during the evolution from reptiles to mammals.

Fish gills were replaced by *lungs,* and the internal surface area of the latter has been greatly increased in mammals in comparison with lower tetrapods by an extensive branching of the respiratory passages within them. A large surface area, combined with efficient ventilation of the lungs by movements of the diaphragm and ribs, provides sufficient gas exchange to sustain a high rate of metabolism. The more cranial respiratory passages have also become modified. With the evolution of a distinct neck, the simple laryngotracheal chamber of amphibians becomes divided into a *larynx,* from which a *trachea* descends to the lungs. And, as explained in connection with the skeleton, the evolution of a bony hard palate and fleshy soft palate separates a respiratory passage from the original oral cavity and cranial part of the pharynx. This permits simultaneous respiration and manipulation of food within the mouth. Food and air passages cross only in the caudal part of the pharynx, and, even here, food is normally prevented from entering the larynx through the evolution of a flaplike *epiglottis.*

Many changes have occurred in the digestive system. Water no longer flows continuously through the mouth in terrestrial vertebrates. In this connection, salivary glands and a tongue began to evolve in amphibians, and both become extensively developed in mammals. The *salivary glands* provide a lubricating secretion and also secrete *ptyalin,* an enzyme that initiates the hydrolysis of starch.

The *tongue,* which aids in manipulating food in the mouth and in swallowing, is a very muscular organ in mammals. The muscles are somatic muscles of hypobranchial origin and, hence, are innervated by the hypoglossal nerve, but the epithelium of the tongue is derived from the pharynx lining, so its sense organs are supplied by certain branchiomeric nerves—general sensory fibers by the trigeminal and taste buds by the facial and glossopharyngeal nerves.

The most notable change in the intestinal region is a great increase in internal surface area. This is accomplished in part through an increase in the length of the intestine, and in part through the evolution of numerous, minute, finger-like *villi.* Another major change is the division of the cloaca into dorsal and ventral portions. The ventral portion contributes to the urogenital passages; the dorsal portion forms the *rectum.*

DIGESTIVE AND RESPIRATORY ORGANS OF THE HEAD AND NECK

(A) SALIVARY GLANDS

The salivary glands of your cat may be studied on the same side of the head that was used for the dissection of the muscles, provided the glands were not injured. If they were destroyed, carefully remove the skin on the opposite side of the head overlying the cheek, throat, and side of the neck ventral to the auricle. The superficial cutaneous muscles (platysma and facial muscles, p. 44) must also be taken off, but be especially careful in the cheek region, for one of the salivary ducts is very superficial.

Pick away the connective tissue ventral to the auricle and you will expose the large, oval-shaped *parotid gland* (Fig. 7–5). It can be recognized by its lobulated texture. The *parotid duct* emerges from the front of the gland, crosses the large cheek muscle (masseter), and perforates the mucous membrane of the upper lip opposite the last premolar tooth. Frequently, accessory bits of glandular tissues are found along the duct. Two branches of the facial nerve going to facial muscles emerge from beneath the

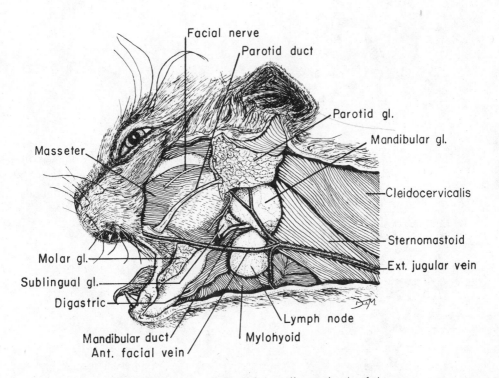

Figure 7–5 Lateral view of the salivary glands of the cat.

parotid gland and cross the masseter, one dorsal and one ventral to the parotid duct. Do not confuse the duct with these. A *mandibular gland* (submandibular gland in human anatomy) lies caudal to the angular process of the jaw and deep to the ventral border of the parotid. It is a large oval-shaped gland having the same lobulated texture as the parotid. Do not confuse it with the smaller, smoother textured *lymph nodes* in this region. The *mandibular duct* emerges from the front of the gland and passes forward, first going lateral to the digastric muscle, which is the large muscle arising from the base of the skull and inserting along the ventral border of the lower jaw (p. 78). Cut and reflect the digastric, and you can see that the duct then passes medial to the caudal border of the mylohyoid — the thin transverse sheet of muscle lying between the paired digastric muscles. Cut and reflect the mylohyoid, and follow the duct forward as far as you can. It is crossed rostrally by the *lingual nerve,* a branch of the trigeminal nerve returning general sensory fibers from the tongue. The *hypoglossal nerve* carrying motor fibers to the tongue musculature lies caudal and dorsal to the mandibular duct. The mandibular ducts of opposite sides converge and enter the floor of the mouth by a pair of inconspicuous openings situated on flattened papillae just rostral to the midventral septum of the tongue, the *lingual frenulum.*

A small, elongated *sublingual gland* is located beside the caudal one-third of the mandibular duct, and generally abuts against the mandibular gland. It is drained by a minute duct which parallels the mandibular duct but is hard to distinguish grossly. The ducts enter the floor of the mouth.

A *zygomatic gland* lies beneath the eye of the cat and has been seen if the eye was

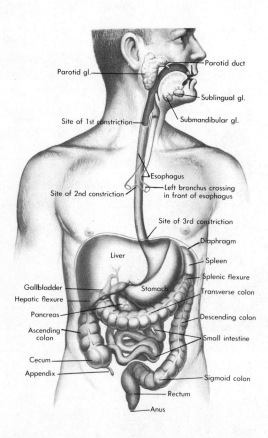

Figure 7–6 The human digestive tract. (From Gardner and Osburn: Structure of the Human Body. W.B. Saunders Company.)

dissected (p. 85). A small, elongated *molar gland* is situated between the skin and mucous membrane of the caudal half of the lower lip. Several small ducts, which cannot be seen grossly, lead from the molar gland to the inside of the lip.

The parotid, submandibular, and sublingual glands are very similar in human beings (Fig. 7–6). The parotid duct opens into the mouth opposite the upper second molar tooth; the submandibular duct terminates beside the lingual frenulum but not on a flap of tissue. We lack the other salivary glands seen in the cat.

(B) ORAL CAVITY

Open the *oral cavity* by cutting through the floor of the mouth with a scalpel. Do this from the external surface, cut on each side, and keep as close to the mandible and chin as possible. Then cut through the symphysis of the mandible with bone scissors, spread the two halves of the lower jaw apart, and pull the tongue ventrally. The rostral part of the roof of the oral cavity is formed by the *hard palate;* the caudal part, by the fleshy *soft palate* (Fig. 7–7 to 7–9). Cats do not have the little tab of flesh, the uvula, that hangs down from the caudal border of our soft palate. A pair of small openings will be seen at the very front of the hard palate just caudal to the incisor teeth. These are the openings of the *incisive ducts.* These ducts pass through the palatine fissure of the skull to the vomeronasal organs in the nasal cavities (p. 87). These organs, which are found in most terrestrial vertebrates, are accessory olfactory organs and they may enable the animal to smell food within the mouth. Traces of them have been found in human embryos, but they disappear later in development.

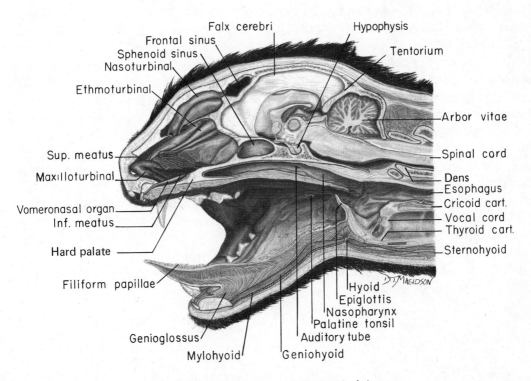

Figure 7–7 Sagittal section of the head of the cat.

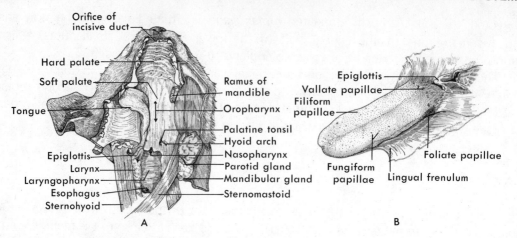

Figure 7–8 Oral cavity and pharynx of the cat. *A*, the floor of the mouth and pharynx have been swung open to the specimen's right; *B,* dorsolateral view of the tongue.

The lateral walls of the oral cavity are bounded by the teeth, lips, and cheeks. That portion of the cavity lying between the teeth and cheeks is called the *vestibule.* A well developed, muscular *tongue (lingua)* lies in the floor of the cavity and is connected to the floor by the vertical lingual frenulum previously seen.

Pull the tongue ventrally sufficiently far to tighten, and bring into prominence, a pair of lateral folds that extend from the sides of the caudal portion of the tongue to the soft palate. These folds constitute the *palatoglossal arches,* and they represent the boundary between the adult oral cavity and *pharynx.* However, part of the back of the oral cavity as thus defined probably develops from the embryonic pharynx. The passage between the palatoglossal arches is called the *fauces.* Notice that the very back of the tongue lies within the pharynx.

Cut through the palatoglossal arch on one side, pull the tongue down more, and examine its dorsal surface (Fig. 7–8). The dorsum of the tongue is covered with papillae, the most numerous of which are the pointed *filiform papillae.* The anterior ones bear spiny projections with which the cat grooms its fur or rasps flesh from bones, but the posterior ones are soft. Small, rounder *fungiform papillae* are interspersed among the filiform, especially along the margins of the tongue. You may have to use a hand lens to see them. Several *vallate papillae* are located near the back of the tongue. Each papilla is a relatively large, round patch set off from the rest of the tongue by a circular groove. There are four to six distributed in a V-shaped line. The apex of the V is directed caudad. Leaf-shaped *foliate papillae* can be found along the side of the tongue lateral to the vallate papillae. Microscopic taste buds are found on the sides and base of all the papillae, except for most of the filiform papillae. We lack foliate papillae, but the others are present. Our filiform papillae are not spiny, and our vallate papillae are more numerous.

(C) PHARYNX

With a pair of scissors, cut caudally through the lateral wall of the pharynx on the side on which the palatoglossal arch was cut. Follow the contour of the tongue to the

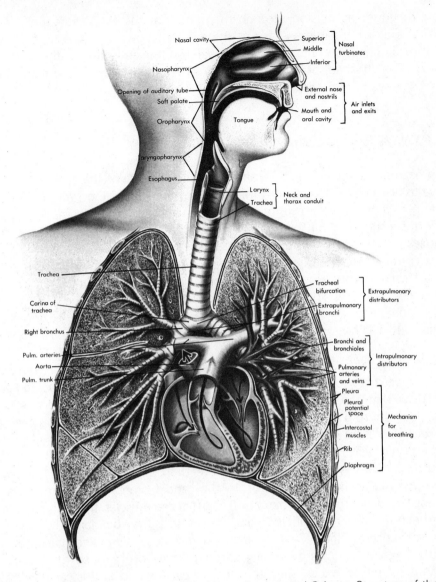

Figure 7-9 The human respiratory system. (From Gardner and Osburn: Structure of the Human Body. W.B. Saunders Company.)

laryngeal region, then extend the cut dorsal to the larynx and caudad into the esophagus. Do not cut into the soft palate or larynx. Swing open the floor of the mouth and pharynx (Fig. 7-8). The pharynx may be divided somewhat arbitrarily into oral, nasal, and laryngeal portions. The *oropharynx* lies between the palatoglossal arches and the free caudal margin of the soft palate. A pair of *palatine tonsils* lies in its lateral walls. Note that each is partially imbedded in a *tonsillar fossa.* The *laryngopharynx* is the space dorsal to the enlargement, the *larynx,* in the floor of the caudal part of the pharynx. It communicates caudally with the *esophagus,* and ventrally with the larynx. The slitlike opening within the larynx is termed the *glottis.* A trough-shaped fold, the *epiglottis,* lies cranial to the glottis and helps to deflect food around or over the glottis. The *nasopharynx* lies dorsal to the soft palate. Open it by making a longitudinal incision through the middle of the soft palate. Spread open the incision as wide as possible and try to shine a light down into the nasopharynx. The pair of slitlike openings in the laterodorsal walls are the entrances of the *auditory tubes* (Fig. 7–7). The choanae, or internal nostrils, enter the rostral end of the nasopharynx but cannot be seen in this view.

(D) LARYNX, TRACHEA, AND ESOPHAGUS

Approach the laryngeal region from the ventral surface of the neck. Several muscles will have to be removed, but do not injure any of the larger blood vessels. The *larynx* is the chamber whose walls are supported by relatively large cartilages. The *hyoid bone,* which forms a sort of sling for the support of the base of the tongue, is embedded in the muscles cranial to the larynx. Its parts have been described elsewhere (p. 28). The larynx is continued caudally as the windpipe, or *trachea,* whose walls are supported by a series of cartilaginous rings. Actually, these rings are incomplete, for their ends do not quite meet dorsally. The *esophagus* is a collapsed, muscular tube lying dorsal to the trachea. Food being swallowed can push the esophagus open as it moves down by peristalsis, but it is necessary for the trachea to be held open for the free movement of air.

The dark *thyroid gland* of the cat lies against the cranial end of the trachea. It consists of two *lobes,* one on either side of the trachea, that are connected across the ventral surface of the trachea by a very narrow band of thyroid tissue called the *isthmus.* Our thyroid lobes extend forward lateral to the larynx. Two pairs of *parathyroid glands* are embedded in the dorsomedial surface of the thyroid, but they cannot be seen grossly.

Return to the larynx and study it more thoroughly. Much of it is covered by *intrinsic laryngeal muscles* derived from the caudal branchiomeric musculature and hence innervated by the vagus nerve. Strip off these muscles from all surfaces of the larynx to expose the laryngeal cartilages. The muscles may be identified as you remove them by reference to Figure 7–10. They are named according to the cartilages between which they extend. The large cranial cartilage that forms much of the ventral and lateral walls of the larynx is called the *thyroid cartilage.* It is this cartilage that forms the projection in the neck of man known as Adam's apple. The ring caudal to the thyroid cartilage is the *cricoid cartilage.* The cricoid is shaped like a signet ring, for its dorsal portion is greatly expanded and forms most of the dorsal wall of the larynx. Careful dissection will reveal a pair of small, triangular cartilages cranial to the dorsal part of the cricoid. These are the *arytenoid cartilages.* Additional, minute cartilages are frequently associated with the arytenoids but are seldom seen. An *epiglottic cartilage* supports the epiglottis.

Cut open the larynx along its middorsal line. The pair of whitish, lateral folds that extend from the arytenoids to the thyroid cartilage are the *vocal cords.* They are set in vibration by the movement of air across them and are controlled by the movement of the arytenoids, the action of muscles within them, and slight changes in the shape of the larynx. The *glottis* is the space between them. The cat has an accessory pair of folds, sometimes called the *false vocal cords,* that extend from the arytenoids to the base of the epiglottis.

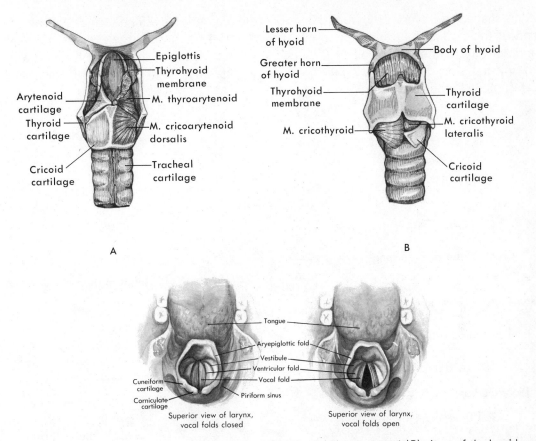

Figure 7–10 The mammalian larynx. Top row, dorsal (*A*) and ventral (*B*) views of the hyoid and larynx of the mink. Intrinsic laryngeal muscles have been removed from the specimen's left side. Bottom row, superior views of the human larynx and vocal cords. (Bottom row from Gardner and Osburn: Structure of the Human Body. W.B. Saunders Company.)

Most of the laryngeal cartilages develop from certain of the visceral arches, but there is some doubt as to the precise homologies. The arytenoids and cricoid are the first to appear phylogenetically. They appear to develop from the sixth, or from the sixth and seventh, visceral arches. The fourth and fifth visceral arches are incorporated in the hyoid apparatus in amphibians. In mammals, the hyoid apparatus involves only the second and third visceral arches, and in most mammals the fourth and fifth arches form the newly evolved thyroid cartilage. Some believe that the tracheal rings may evolve from a splitting and multiplication of the seventh arch, but this is doubtful. The epiglottic cartilage is apparently a new structure.

THORAX AND ITS CONTENTS

(A) PLEURAL CAVITIES

Open the thorax by making a longitudinal incision about two centimeters to the right of the midventral line, and extending the length of the sternum. Use a strong pair of scissors. Spread open the incision and look into the *right pleural cavity*. A dome-shaped, transverse, muscular partition (the *diaphragm*) will be seen at the caudal end of the cavity. Make another cut just cranial to the diaphragm that extends laterally and dorsally to the back. Follow the line of attachment of the diaphragm but keep on the pleural side. Spread the right thoracic wall laterally, breaking the ribs near their attachment to the vertebrae.

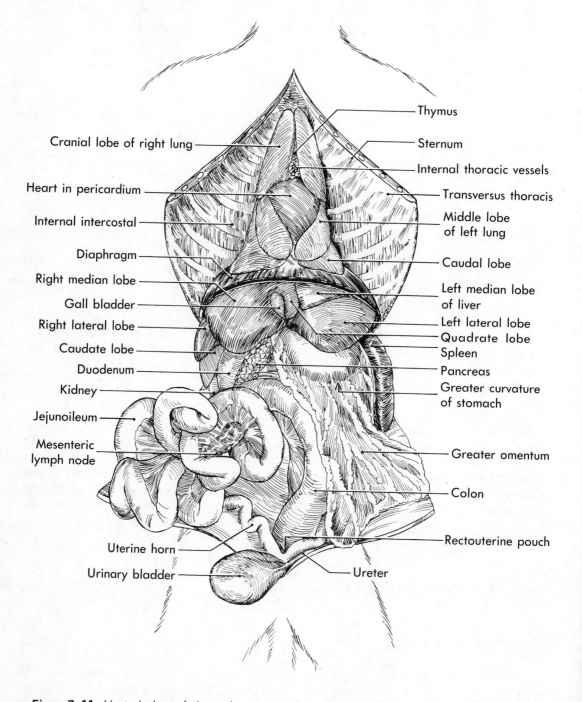

Figure 7–11 Ventral view of the major organs of the thoracic and peritoneal cavities of a cat.

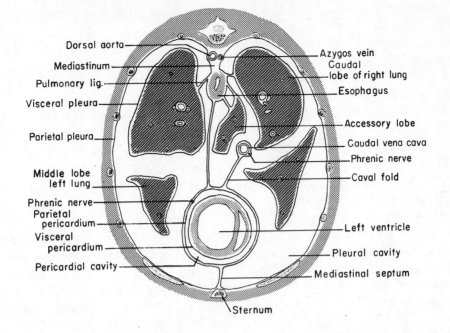

Figure 7–12 A diagrammatic transverse section through the thorax of the cat at the level of the ventricles to show the coelomic epithelium and its relation to the thoracic viscera. The section is viewed from behind so that left and right sides of the animal and drawing coincide.

The right pleural cavity and its *lung (pulmo)* are now well exposed (Fig. 7–11). The coelomic epithelium lining the walls of the pleural cavity is called the *parietal pleura;* that covering the surface of the lung, the *visceral (pulmonary) pleura.* The right lung of the cat is divided into four *lobes:* cranial, middle, caudal and accessory. The accessory lobe extends dorsal to a large vein *(caudal vena cava)* and then ventrally into a pocket on the medial side of the mesentery *(caval fold)* attaching to the ventral surface of the vein (Fig. 7–12). Tear the caval fold near the vein and you can see this lobe. It, together with the caval fold, are absent in human beings. The lobes of the lung are attached to the medial wall of the pleural cavity by a pleural fold known as the *pulmonary ligament.* The blood vessels connected with the heart, and the *bronchi* from the trachea to the lung, pass through part of this ligament, but they should not be dissected at this time. These structures constitute the *root of the lung.* Cut into a part of the lung and notice that it is not an empty organ, but a very spongy one. The numerous, thin-walled, terminal pockets of the respiratory passages *(alveoli),* in which gas exchange occurs, are not visible grossly.

The medial wall of each pleural cavity (right and left) is formed only of a layer of parietal pleura (Fig. 7-12). The space, or potential space, between the medial walls of the two pleural cavities constitutes the *mediastinum.* This space, however, is largely filled with structures that lie between the two cavities. For example, the pericardial cavity and heart, which form the large bulge medial and ventral to the lung, lie in the mediastinum. In places the medial walls of the pleural cavities meet and form a mesentery-like structure termed the *mediastinal septum.* The mediastinal septum can be seen caudal to the heart and medial to the accessory lobe of the right lung. The caval fold is an evagination from this portion of the septum. In the cat the mediastinal septum continues craniad ventral to the heart.

Break the line of attachment of the parietal pleura to the ventral thoracic wall. You will have to cut some blood vessels passing to the ventral thoracic wall, but do so in such a way that their ends can be apposed later. Cut laterally and dorsally through the left thoracic wall close to the diaphragm in the same way that you did on the right. Turn back the body wall and examine the *left pleural cavity.* (This will give a good view of the third muscle layer of the thorax, transversus thoracis, referred to on page 73.) The left lung of the cat does not have an accessory lobe; in human beings it is divided only into inferior (caudal) and superior (cranial) lobes. The left lung of mammals is always slightly smaller than the right one because of the displacement of the heart toward the left side of the body.

Pull the left lung ventrally and examine the region dorsal to it. A large artery, the *aorta,* and the *esophagus* can be seen passing through the dorsal portion of the mediastinum. The aorta lies to the left of the vertebral column; the esophagus, more ventrally. Let the lungs fall back into place. A pair of white strands, the *phrenic nerves,* can be seen in the central portion of the mediastinum on each side of the pericardial cavity and heart. They lie ventral to the roots of the lung and pass caudad to the diaphragm. The right one follows the caudal vena cava closely; the left passes through the caudal part of the mediastinal septum. The origin of these nerves from the ventral rami of the fifth and sixth cervical nerves is indicative of the cervical derivation of the diaphragmatic muscles. The portion of the mediastinum ventral and cranial to the heart is occupied by the dark, irregularly lobulated *thymus.* The thymus varies considerably is size, being best developed in young individuals.

(B) PERICARDIAL CAVITY

The *pericardial cavity* and *heart* are the largest structures within the mediastinum. Cut open the pericardial cavity by a midventral incision. Its wall, known as the *pericardium,* is formed of connective tissue and coelomic epithelium, the *parietal pericardium. Visceral pericardium* covers the surface of the heart. Parietal and visceral pericardium are continuous with each other over the vessels at the cranial end of the heart.

PERITONEAL CAVITY AND ITS CONTENTS

(A) BODY WALL AND PERITONEAL CAVITY

Make a longitudinal incision through the abdominal wall slightly to the right of the midventral line. Extend the cut from the diaphragm to the pelvic girdle. Then cut laterally and dorsally along the attachment of the diaphragm to the body wall. Do this on both sides, thereby freeing the diaphragm as far as the back. Reflect the flaps of the abdominal wall. Notice that, apart from the skin, you have cut through three layers of muscle or their aponeuroses (external oblique, internal oblique, and transversus abdominis), possibly the rectus abdominis near the midventral line, and the parietal peritoneum. The portion of the coelom exposed is the *peritoneal cavity.* Its walls are lined with *parietal peritoneum,* and its viscera covered with *visceral peritoneum.* Wash out the cavity if necessary.

(B) ABDOMINAL VISCERA AND MESENTERIES

. The concave surface of the dome-shaped diaphragm forms the cranial wall of the peritoneal cavity, and the large *liver (hepar)* lies just caudal to it and is shaped to fit into the dome. Pull the liver and diaphragm apart. It can now be seen that the central portion of the diaphragm is formed by a tendon (the *central tendon*), into which its muscle fibers insert. A vertical *falciform ligament* extends between the diaphragm, liver, and ventral wall. Sometimes a thickening, which represents a vestige of the embryonic umbilical vein, may be seen in its free edge. It is known as the *round ligament* of the liver. Diaphragm and liver are closely apposed dorsal to the falciform ligament, and the reflections of the peritoneum from the one to the other in this region constitute the *coronary ligament.*

The liver can be divided into right and left halves at the cleft into which the falciform ligament passes. Each half is divided into a lateral and a medial lobe, thus making a *left lateral, left medial, right medial,* and *right lateral lobe* (Fig. 7–11). A small *quadrate lobe* is interposed between the left medial and right medial lobes. It is partly united with the latter, being separated from it by the *gall bladder (vesica fellea).* A *caudate lobe* lies caudal to the right lateral lobe and abuts against the right kidney. Part of the caudate lobe extends toward the left side of the body going deep to a mesentery, the *lesser omentum,* which extends from the liver to the stomach and duodenum. A *hepatorenal ligament* extends from this lobe to the parietal peritoneum near the right kidney. The left kidney lies in a slightly more caudal position on the opposite side of the body. The human liver is subdivided in a similar way.

As in the majority of vertebrates, most of the *stomach (ventriculus)* lies on the left side of the peritoneal cavity, and the organ is more or less J-shaped (Figs. 7–6 and 7–11). Cut through the left side of the diaphragm to find the point at which the *esophagus* enters. There is an abrupt change in the diameter of the digestive tract at this point. The portion of the stomach adjacent to the esophagus is its *cardiac region;* the dome-shaped portion extending cranially to the left of the cardiac region, its *fundus;* the main part of the stomach, its *corpus;* and the narrow caudal portion, its *pyloric region,* but these gross regions do not necessarily correspond with glandular regions bearing the same names. The stomach ends in a thick muscular sphincter, the *pylorus,* which can be seen if you cut open the stomach. Do not injure any mesenteries. You will also see longitudinal ridges in the lining that are called *rugae.* The long left and caudal margin of the stomach, which represents its original dorsal surface, constitutes its *greater curvature;* the shorter right and cranial margin, which represents its original ventral surface, the *lesser curvature.*

Notice that the lesser omentum, which represents a part of the ventral mesentery, attaches along the lesser curvature of the stomach. The mesentery that attaches along the greater curvature is the *greater omentum* or mesogaster—a part of the dorsal mesentery. The greater omentum of mammals does not extend directly to the middorsal line of the peritoneal cavity, as it does in lower vertebrates, but is modified to form a saclike structure (the *omental bursa*), which drapes down over the intestines (Fig. 7–11). The omental bursa is very large, contains considerable fat in its wall, and extends over the intestine nearly to the pelvic region. It is often entwined with the intestine and must be untangled carefully. Observe that the greater omentum extends caudally from its line of attachment on the stomach, and then turns upon itself and extends cranially and dorsally to attach onto the dorsal wall of the peritoneal cavity. The *spleen (lien)*, which is relatively much larger in the cat than in human beings, lies in the wall of the omentum on the left side of the stomach. That portion of the greater omentum between the spleen and stomach is known as the *gastrolienic ligament.* A small, triangular mesentery, the *gastrocolic ligament,* passes from the part of the greater omentum lying dorsal to the spleen over to the mesentery of the large intestine.

A part of the peritoneal cavity known as the *lesser peritoneal cavity* lies between the descending and ascending walls of the omental bursa. Tear the bursa and verify that it contains a space. At one stage of development, the lesser peritoneal cavity has a wide communication with the main part of the peritoneal cavity (Fig. 7–13). But subsequent adhesions of the liver to the dorsal part of the diaphragm and adjacent body wall reduce this to a relatively small *epiploic foramen* (Fig. 7–14). The epiploic foramen lies dorsal to the lesser omentum, and between the caudate lobe of the liver and the mesentery to the duodenum. If your specimen is large enough, you can pass a finger through the epiploic foramen and extend it dorsal to the stomach and into the omental bursa.

Carefully dissect that portion of the lesser omentum lying near the caudate lobe of the liver and the epiploic foramen in order to expose the system of bile ducts extending from the liver and gall bladder to the beginning of the duodenum. Some lymphatic vessels, which look like chains of small nodules, may have to be removed. A *cystic duct* comes down from the gall bladder and unites with several *hepatic ducts* from various parts of the liver to form a *common bile duct (ductus choledochus)* which passes to the duodenum. One particularly prominent hepatic duct comes in from the left lobes of the liver, and another from the right lateral lobe.

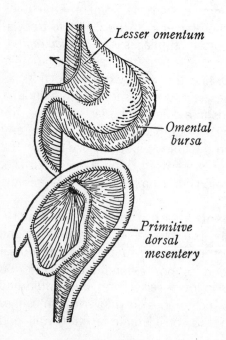

Lesser omentum

Omental bursa

Primitive dorsal mesentery

Figure 7–13 Diagrammatic ventral view of the mesenteries of a mammalian embryo to show the formation of the omental bursa from the posterior extension of the mesogaster. The liver would attach to the lesser omentum along its cut edge (double line). The arrow indicates the approximate position of the epiploic foramen. (From Arey: Developmental Anatomy.)

The *small intestine* extends caudad from the stomach, passes through numerous convolutions, and eventually enters a *large intestine* (Figs. 7–6 and 7–11). The small intestine of mammals has differentiated into a cranial *duodenum* and a more caudal *jejunum* and *ileum.* The duodenum is the first, approximately U-shaped, loop of the intestine. It curves caudad from the pylorus on the right side of the body, and then bends toward the left and ascends nearly to the stomach. It is arbitrarily considered to end at its next major bend. The duodenum is 15 to 20 centimeters long in an adult cat. A small, triangular-shaped *duodenorenal ligament* passes from the mesoduodenum to the dorsal wall medial to the right kidney. There is no sharp transition between jejunum and ileum. About all that can be said in respect to their gross anatomy is that the jejunum is the cranial half of the postduodenal small intestine, and the ileum the caudal half. The small intestine is supported by a part of the dorsal mesentery—that portion passing to the duodenum being the *mesoduodenum,* and that portion to the jejunum and ileum, the *mesentery.*

Cut open a part of the small intestine, and examine its lumen. The lining has a velvety appearance which results from the presence of numerous, minute, finger-like projections called *villi.* These greatly increase the internal surface area. Parasitic roundworms and tapeworms are often found in the intestine of the cat.

The *pancreas* can be recognized by its lobulated texture. A part of it lies against the descending portion of the duodenum, and a part of it extends as a taillike process transversely across the body to the spleen. This portion lies in the dorsal wall of the omental bursa. Two pancreatic ducts are present, for the ducts of both the dorsal and the ventral primordium are retained. The main duct, known as the *pancreatic duct,* unites with the common bile duct as the latter enters the duodenum. It can be found by carefully picking away pancreatic tissue in this region. The enlargment on the duodenum where the two unite is known as the *hepatopancreatic ampulla* (Fig. 7–14). An *accessory pancreatic duct,* which would be the duct of the dorsal pancreas, enters the duodenum about one centimeter caudal to the main duct, but it is small and very hard to find. Our bile and pancreatic ducts are essentially the same.

Follow the coils of the rest of the small intestine (jejunum and ileum) until it enters the much wider colon. A short, blind diverticulum, the *cecum,* extends caudally from the beginning of the colon. A small, blind tube, the vermiform appendix, extends from the human cecum (Fig. 7–6), but it is absent in the cat. Cut open the wall of the cecum and colon opposite the entrance of the ileum. Notice that the ileum projects slightly into the lumen of the colon, forming an *ileal papilla,* which helps prevent the backing up of colic material into the small intestine.

The large intestine of mammals is much longer than that of lower vertebrates and has a greater diameter than the small intestine. Most of it constitutes the *colon,* and it is supported by a portion of the dorsal mesentery termed the *mesocolon.* The colon itself extends forward on the right side of the body for a short distance *(ascending colon),* crosses to the left side *(transverse colon),* and extends back into the pelvic canal *(descending colon).* The pattern of the human intestine is quite similar to that of the cat (Fig. 7–6). Our ascending colon is longer, and the caudal end of our colon makes a small S-shaped bend (sigmoid colon) which a cat does not have.

The caudal portion of the colon also lies dorsal to the pear-shaped *urinary bladder* and, if your specimen is a female, to the Y-shaped *uterus* (Fig. 7–11). Notice that the bladder is supported by a vertical *median ligament,* which, being a part of the ventral mesentery, extends to the midventral body wall, and by a pair of *lateral ligaments.* The

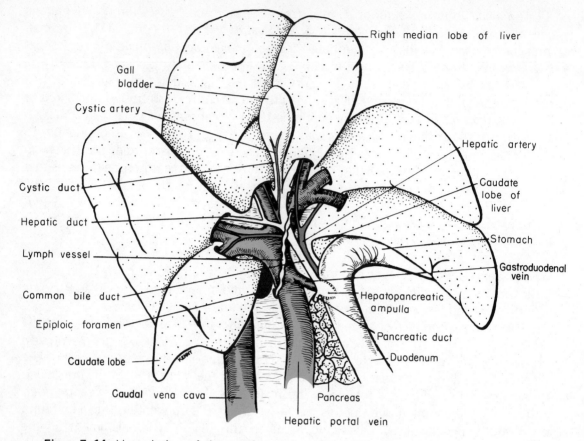

Gall
bladder

Cystic artery

Cystic duct

Hepatic duct

Lymph vessel

Common bile duct

Epiploic foramen

Caudate lobe

Caudal vena cava

Right median lobe of liver

Hepatic artery

Caudate
lobe of
liver

Stomach

Gastroduodenal
vein

Hepatopancreatic
ampulla

Pancreatic duct

Duodenum

Pancreas

Hepatic portal vein

Figure 7–14 Ventral view of the vessels and ducts in the lesser omentum in the cat. The right median and part of the right lateral lobe of the liver have been turned forward.

latter often contain wads of fat. Cut open part of the colon, clean it out, and notice that it lacks villi. Also notice the extension of the coelom into the pelvic canal. That portion of the coelom in the male that extends caudally between the large intestine and the urinary bladder is called the **rectovesical pouch.** The comparable coelomic extension in the female is divided by the uterus into a shallow **vesicogenital pouch** between the bladder and uterus, and a deep **rectogenital pouch** between the uterus and large intestine.

Deep within the pelvic canal, the colon passes into the terminal segment of the large intestine, the **rectum,** which, in turn, opens on the body surface through the **anus.** The rectal region will be seen later when the pelvic canal is opened.

Chapter Eight

THE CIRCULATORY SYSTEM

FUNCTIONS OF THE CIRCULATORY SYSTEM

Continuing with the organ systems that provide for the metabolic needs of the body, we will next consider the circulatory system. This system is primarily the great transport system of the body. Oxygen and food are carried from the respiratory and digestive organs to all the tissues and cells; carbon dioxide and other excretory products are carried from the tissues to sites of removal; and hormones are transported from the endocrine glands to the responding tissues. But the system has other functions. It aids in combating disease and in repairing tissues and plays an important role in maintaining the constancy of the internal environment in many ways.

PARTS OF THE CIRCULATORY SYSTEM

The blood and lymph and their cells are functionally the most important parts of the circulatory system, but only the vessels that propel and carry these can be studied in a course of this scope. These vessels may be subdivided into (1) a *cardiovascular system,* consisting of the *heart, arteries, blood capillaries,* and *veins;* and (2) a *lymphatic system,* consisting of closed *lymphatic capillaries* and *lymphatic vessels.* In addition, the lymphatic system of higher vertebrates includes many *lymph nodes* located at strategic junctions along the course of the vessels. These nodes act as sites for the

Figure 8–1 Diagrammatic lateral view of the major blood vessels of a 26 day old human embryo. Most vessels are paired and only those on the left side are shown. (From Moore: The Developing Human. W.B. Saunders Company.)

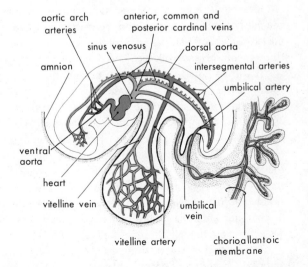

aortic arch arteries

anterior, common and posterior cardinal veins

sinus venosus

dorsal aorta

amnion

intersegmental arteries

umbilical artery

ventral aorta

heart

vitelline vein

umbilical vein

vitelline artery

chorioallantoic membrane

production of certain white blood cells *(lymphocytes)* and cells that transform into antibody-producing cells. The *spleen* is similar to a lymph node, but is interposed in the cardiovascular system. It is, at different periods of development and in different vertebrates, a site for the production, storage, and elimination of blood cells.

Briefly, the course of the circulation through these two systems of vessels is as follows: *Blood* leaves the heart and travels through the arteries to the capillaries in the tissues. At this point, some of the blood plasma leaves the capillaries to circulate among the cells as *tissue fluid.* Much of the tissue fluid re-enters the capillary bed and returns to the heart through veins. However, in all but the most primitive vertebrates, some enters the lymphatic capillaries and is carried as *lymph* by the lymphatics to the larger veins. Because pressure is very low in lymphatic capillaries and their walls are more permeable than those of cardiovascular capillaries, the lymphatic system can also return large molecules in the tissue fluid, including those plasma proteins that escape from the blood.

The heart is the major pump in causing the fluids to circulate, but by the time the blood reaches the veins the pressure is relatively low. Moreover, pressures created by the heart do not directly affect the lymphatic system. Thus the return of lymph, and to some extent the return of blood, are implemented by other forces. Among these, the contraction and tonus of the surrounding body muscles play a major role. The veins and lymphatics also contain *valves* that prevent a back flow of the fluids.

The lymphatic system is rather obscure and not commonly seen in gross dissections. Parts of it will be studied, but the emphasis throughout will be on the more conspicuous cardiovascular system.

DEVELOPMENT AND EVOLUTION OF THE CARDIOVASCULAR SYSTEM

Although the pattern of blood vessels in the adults of the several classes of vertebrates is quite different, all have a very similar pattern in their early embryos. A brief consideration of this early pattern is necessary for an understanding of the adult cardiovascular system. The first vessels to take definite form in the embryo are a pair of *vitelline veins* (Fig. 8–1) that lie beneath the embryonic gut and carry blood and food forward from the yolk-laden archenteron, or yolk sac, if such a sac is present. (Veins are defined as vessels that carry blood toward the heart.) The cranial parts of these vessels fuse to form the *heart* and *ventral aorta.* Caudal to the heart, they are engulfed and broken up into a capillary network by the enlarging liver. The portion of the vitelline veins lying caudal to the liver becomes the *hepatic portal system;* the portion from the liver to the heart, the *hepatic veins.* *(Portal veins* are simply veins that, after draining one capillary bed, pass to another capillary bed in a different organ. Veins going directly to the heart are referred to as *systemic veins.*)

Meanwhile, a series of six aortic arches develop in the first six branchial bars, which lie between the gill pouches, to carry blood from the heart and ventral aorta up to the *dorsal aortae.* The dorsal aortae, which are first paired but later fuse, carry the blood caudally and, by *vitelline arteries,* back to the archenteron and yolk sac. This completes one circuit.

It will be noted that the early circulation is largely a visceral circulation to and from the "inner tube" of the body, but a somatic circulation to the "outer tube" soon appears. Other branches from the dorsal aorta carry blood into the body wall and out to an allantois and placenta (by *umbilical arteries*) if one is present. Blood from the dorsal portions of the body returns by way of *anterior* and *posterior cardinals.* The cardinals of each side unite cranial to the liver, and turn ventrally to the heart as *common cardinals* or ducts of Cuvier. It will be recalled that the common cardinals pass through the transverse septum which they helped to form. Blood from the more lateral and ventral portions of the body wall, and from the allantois and placenta, if present, returns by a pair of more ventrally situated *umbilical veins.* These vessels enter the base of the common cardinals in early embryos, but in the later embryos of the higher vertebrates they acquire a connection with the hepatic portal system and are drained through the liver.

The pattern of vessels in an adult fish is very similar to that just described for an early amniote embryo. Changes that occur in the cardiovascular system during the evolution to mammals correlate for the most part with the transition from water to land and with the increase in activity and metabolism that characterizes mammals. The loss of gills and the evolution of efficient lungs are correlated with the complete division of the simple tubular heart of fishes into one with a right side receiving depleted blood from the body and sending it to the lungs, and a left side receiving aerated blood from the lungs and sending it to the body. Loss of gills and the complete division of the heart have been important factors in greatly increasing blood pressure and the general efficiency of circulation. For example, the mean blood pressure in the dorsal aorta of a dogfish is about 17 mm. Hg whereas that in the caudal artery of a mouse is 136 mm. Hg. Correlated with the loss of gills are extensive modifications in the aortic arches; the first, second, and fifth are lost, and the others are transformed into a group of arteries which leave the heart to supply the lungs, head, and body. The anterior and posterior cardinal systems, which are the major portions of the systemic venous system of a fish and an early mammalian embryo, undergo complex transformations into a cranial and caudal caval system. These changes will be described in connection with the appropriate vessels.

In the development of an embryo, especially embryos of the higher vertebrates which have had a long and complex phylogenetic history, we see a succession of vessels. Much of the variation seen in the vessels of the adult mammal can be attributed to the persistence of embryonic channels that normally atrophy and to the failure of certain later channels to develop. Other variation results from the enlargement of one channel over another in a primordial capillary plexus that exists in many parts of the embryo. The relative rate of blood flow, as well as hereditary and other factors, is an important factor in determining which channels will enlarge.

THE STUDY OF BLOOD VESSELS

The blood vessels should be studied on specimens that are at least doubly injected. Triply injected specimens are necessary to see all of the hepatic portal system, and quadruply injected ones for seeing most of the lymphatic system.

While dissecting the vessels, remember that they are subject to considerable variation and may not be just as described. Since the veins of the higher vertebrates have had a more complex ontogeny than the arteries, and the venous blood has a more sluggish flow, it is to be expected that more variations will be found in the venous system than in the arterial system. Odd as it may at first seem, the peripheral parts of the vessels are subject to less variation than are certain of the more central and larger channels. For example, the left ovarian vein of a mammal always drains the ovary, but it may enter either the left renal vein or the caudal vena cava. Hence, if a vessel cannot be identified from its point of connection with a major vessel, it can be identified if its peripheral distribution can be established.

The blood vessels of vertebrates are exceedingly numerous, and not all can be studied in a course of this scope. Emphasis has been placed on the major channels in the axis of the body and the vessels connecting with them. The blood vessels of the appendages are treated superficially.

MAMMALIAN CIRCULATION

In most parts of the body, the arteries and the veins will be described together because they tend to parallel each other, and therefore it is convenient to dissect them at the same time. Veins are often harder to find than the arteries, for valves frequently prevent the injection mass from reaching the peripheral parts of these vessels, but identification of the accompanying artery will help the student find the vein. If not injected, veins appear as translucent, fluid-filled tubes beside the corresponding arteries.

Before studying the blood vessels on a regional basis, it is desirable to understand the overall

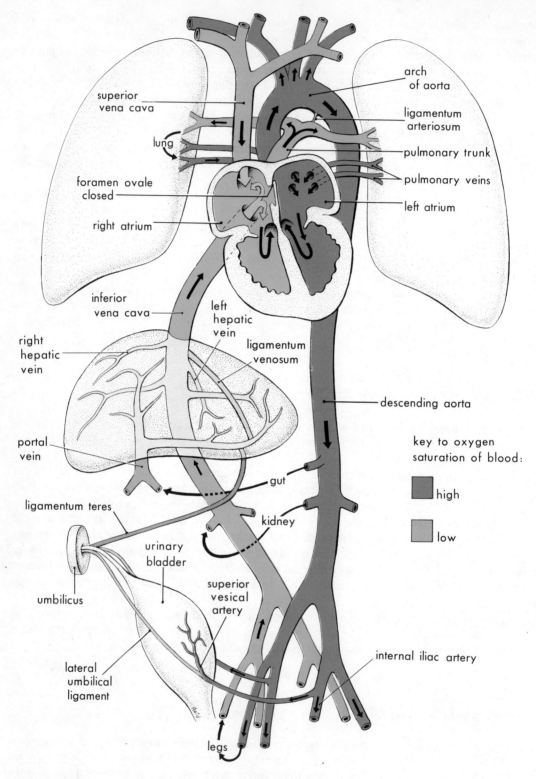

Figure 8–2 Diagram of the adult pattern of circulation in a mammal. The diagram is a ventral view so the right side of the heart and body is on the left side of the diagram. (From Moore: The Developing Human. W.B. Saunders Company.)

aorta

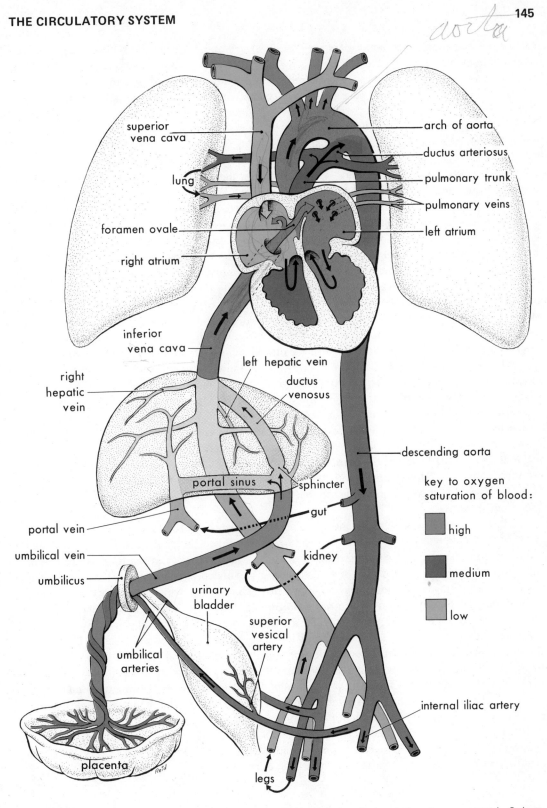

Figure 8–3 Diagram in ventral view of the fetal pattern of circulation in a mammal. Colors indicate approximate oxygen saturation of the blood in the various vessels. (From Moore: The Developing Human. W.B. Saunders Company.)

pattern of circulation. In an adult mammal (Fig. 8–2), tributaries of the *caudal* (inferior) *vena cava* drain the body caudal to the diaphragm, and those of the *cranial* (superior) *vena cava* drain the body cranial to the diaphragm. Both venae cavae return venous blood low in oxygen to the *right atrium* of the heart. From here the blood goes to the *right ventricle,* from which it is pumped to the lungs through the *pulmonary trunk* and *arteries.* After aeration in the lungs, arterial blood returns through the *pulmonary veins* to the *left atrium.* From here it goes to the *left ventricle,* which pumps it out the *aorta.* Branches of the aorta carry arterial blood to nearly all parts of the body. However, blood circulating through the chambers of the heart does not supply the musculature of this organ, so a separate *coronary system* is necessary. As you will see, the coronary arteries leave the base of the aortic arch and veins return to the right atrium via a coronary sinus.

The pattern of the fetal circulation is somewhat different because the placenta, rather than the digestive tract, the kidneys, and the lungs, is the site for the intake of food, the elimination of wastes, and the exchange of gases. An understanding of the fetal pattern is necessary to understand certain functionless remnants of fetal passages that persist in the adult. Blood rich in nutrients and oxygen and low in waste products enters the fetus through the *umbilical vein* (Fig. 8–3). Most of this blood passes directly through the liver in the *ductus venosus* to enter the caudal vena cava, but there is some admixture of venous blood in the liver. Since the entrance of the caudal vena cava into the right atrium is directed toward the interatrial septum, and since blood pressure is relatively low in the left atrium because little blood returns to this chamber from the functionless lungs, most of the rich caudal vena caval blood passes through a valved opening in the interatrial septum, the *foramen ovale,* to enter the left side of the heart. This blood is pumped into the aorta by the left ventricle. Venous blood returning in the cranial vena cava, along with some admixture of blood from the caudal vena cava, passes through the right atrium into the right ventricle. This blood, which is not as rich as the blood in the left side of the heart, is pumped toward the lungs by the right ventricle. Some of it goes to the lungs, but, because the lungs are collapsed and offer considerable resistance to blood flow, most of it by-passes the lungs via a *ductus arteriosus* and enters the aorta distal to the origin of the major arteries to the head and the arms. The locations of the ductus arteriosus and the foramen ovale are such that the head, the brain, and the upper part of the trunk of the fetus receive the richer blood, and a highly mixed blood is distributed to the rest of the body and, via the umbilical arteries, back to the placenta.

As development proceeds, more and more blood flows through the lungs. At birth, the lungs fill with air, and pulmonary resistance to blood flow is less than that of the rest of the body. Blood returning from the lungs increases pressure in the left atrium, the valve in the foramen ovale is held shut, and all the blood in the right side of the heart is pumped to the lungs. As time goes on, the foramen ovale permanently closes, leaving only a depression in this region, the *fossa ovalis.* Blood in the left side of the heart, all of which has been through the lungs, is pumped into the aorta. Most blood continues to the body, but some in the aorta, because of the relatively low pulmonary resistance, flows through the ductus arteriosus into the lungs. This temporary reversed flow through the ductus arteriosus gives a certain fraction of blood a double aeration in the lungs. Within a few hours, the ductus arteriosus contracts. Eventually its lumen fills in with connective tissue, and the duct is transformed into the *ligamentum arteriosum.* With the loss of the placenta, the umbilical vein and arteries lose their function. The vein becomes the *round ligament* (ligamentum teres) of the liver, which may have been seen in the falciform ligament (p. 137). The proximal portions of the umbilical arteries remain as the proximal portions of the *internal iliac arteries,* but their distal parts atrophy and become *lateral umbilical ligaments,* which sometimes can be seen on either side of the urinary bladder. The adult circulatory pattern has now been established.

HEART AND ASSOCIATED VESSELS

Carefully cut away the pericardial sac and thymus from around the heart and its great vessels. The *heart (cor)* is a large, compact organ having a pointed caudal end (its *apex*) and a somewhat flatter cranial surface (its *base*). The *right* and *left ventricles,* which are completely separated internally, form the caudal two-thirds or more of the organ. They are approximately conical in shape and have thick, muscular walls (Fig. 8–16). The *right* and *left atria,* also completely separated internally, lie cranial to the ventricles, and are set off from them by a deep, often fat-filled groove called the *coronary sulcus.* The atria are thinner-walled and darker than the ventricles. They are separated from each other on the ventral surface by the great arteries leaving the cranial end of the ventricles. That portion of each atrium lying lateral to these arteries is called the *auricle.* The auricles are somewhat ear-shaped and tend to have scalloped margins. The separation between the ventricles appears, on the ventral surface, as a shallow groove extending from the left auricle diagonally and toward the right.

Pick away the fat from around the large arteries, leaving the cranial end of the ventricles (Fig. 8–4). The more ventral vessel is the *pulmonary trunk.* It arises from the right ventricle and extends dorsally. Trace it later. The more dorsal vessel is the *arch of the aorta.* It arises from the left ventricle deep to the pulmonary trunk, but one cannot see much of it until it emerges on the right side of the pulmonary trunk. As you continue to pick away fat and loose connective tissue from around these vessels, you will notice that they are bound together by a tough band of connective tissue, the *ligamentum arteriosum,* which is the remnant of the embryonic ductus arteriosus. Try not to destroy it. Just after this connection, the pulmonary trunk bifurcates and its branches, the *pulmonary arteries,* pass to the left and right lungs. This bifurcation is most easily seen by pushing the pulmonary trunk cranially and dissecting between it and the craniodorsal portion of the heart. Two small *coronary arteries* leave the base of the arch of the aorta and pass to the heart wall. One can be found deep between the pulmonary trunk and the left auricle; the other, deep between the pulmonary trunk and the right auricle (Fig. 8–5). *Coronary veins* draining the heart wall parallel the arteries.

Push the heart to the left side of the thorax, and you will see the *caudal* (inferior) *vena cava,* or postcava, coming through the diaphragm and entering the right atrium. A *cranial* (superior) *vena cava,* or precava, will also be seen entering this chamber from the right side of the neck. The coronary veins collect into a *coronary sinus* which enters the right atrium almost in common with the caudal vena cava. This vessel lies on the dorsal surface of the heart between the atria and ventricles and can be seen by lifting the apex of the heart. Carefully pick away connective tissue dorsal to the cranial vena cava and from the roots of the lungs. You will find the *pulmonary veins* coming from the lungs and entering the left atrium. There are several veins, but those of each side generally collect into two channels before entering the heart.

ARTERIES AND VEINS CRANIAL TO THE HEART

(A) VESSELS OF THE CHEST, SHOULDER, ARM, AND NECK

Trace the *cranial vena cava* forward and carefully expose its tributaries by picking away surrounding portions of the thymus, connective tissue, and fat (Fig. 8–4). A *subclavian vein* comes in from each shoulder and arm just in front of the first rib (a valve often prevents it from being injected) and joins a *bijugular trunk,* which receives

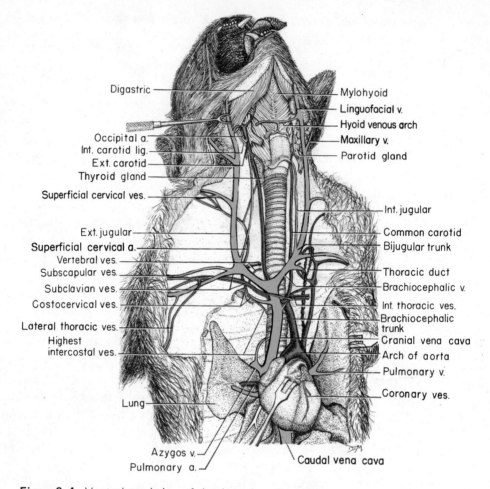

Digastric — — Mylohyoid
— Linguofacial v.
— Hyoid venous arch
Occipital a. — — Maxillary v.
Int. carotid lig. —
Ext. carotid — — Parotid gland
Thyroid gland —
Superficial cervical ves. —
— Int. jugular
Ext. jugular — — Common carotid
Superficial cervical a. — — Bijugular trunk
Vertebral ves. —
Subscapular ves. — — Thoracic duct
Subclavian ves. — — Brachiocephalic v.
Costocervical ves. — — Int. thoracic ves.
— Brachiocephalic trunk
Lateral thoracic ves. — — Cranial vena cava
Highest intercostal ves. — — Arch of aorta
— Pulmonary v.
— Coronary ves.
Lung —
Azygos v. —
Pulmonary a. — — Caudal vena cava

Figure 8–4 Ventrolateral view of the thoracic and cervical arteries and veins of the cat.

the drainage from one side of the neck and head. This union forms the *brachiocephalic veins.* Left and right brachiocephalics, in turn, unite to form the cranial vena cava.

Next examine certain of the tributaries of the vena cava and brachiocephalic. The most caudal tributary, entering the dorsal surface of the vena cava, is the *azygos vein,* which receives most of the *intercostal veins* from between the ribs on both sides of the body. *Intercostal arteries* will be seen beside the veins; their origins will be seen later. A *highest intercostal vein,* which drains the cranial intercostal spaces, enters the azygos.

The next cranial tributaries are several small veins from the thymus and a larger *internal thoracic vein,* which enters the ventral surface of the cranial vena cava. At its entrance it is a single vessel, but it bifurcates distally and drains both sides of the ventral thoracic wall. Its distal parts lie deep to the transversus thoracis muscle and are accompanied by the *internal thoracic arteries,* whose origin will be seen soon. The internal thoracic vessels continue into the cranial part of the ventral abdominal wall where they are called the *cranial epigastric arteries* and *veins.*

Return to the arch of the aorta. After giving off the coronary arteries previously

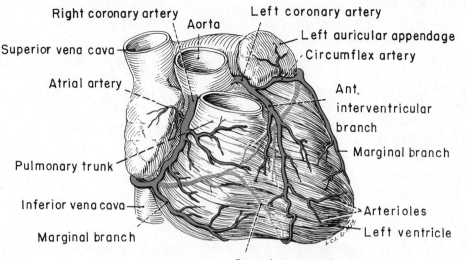

Right coronary artery   Left coronary artery

Aorta

Superior vena cava — Left auricular appendage

Circumflex artery

Atrial artery

Ant.
interventricular
branch

Marginal branch

Pulmonary trunk

Inferior vena cava — Arterioles

Left ventricle

Marginal branch

Post. interventricular branch

Figure 8–5 Ventral view of the human heart and coronary arteries. (From King and Showers: Human Anatomy and Physiology, W.B. Saunders Company.)

described, the aorta curves dorsally and to the left, disappearing dorsal to the root of the left lung. Two vessels arise from the front of the arch—a large *brachiocephalic trunk* nearest the heart and then a smaller, *left subclavian artery.* Trace the brachiocephalic forward. It sends off small branches to the thymus, and then breaks up into three vessels—two *common carotid arteries* that ascend the neck on either side of the trachea, and a *right subclavian artery.* These vessels continue cranially deep to the vena cava and brachiocephalic veins.

Trace one of the subclavian arteries peripherally, preferably on the side of the body in which the veins are well injected so that they can also be identified. Medial to the first rib, the subclavian artery gives rise to four branches which are most accurately identified from their peripheral distribution. The *internal thoracic artery,* previously identified, leaves the ventral surface of the subclavian and accompanies the internal thoracic vein to the ventral chest wall.

A *vertebral artery* arises from the dorsal surface of the subclavian nearly opposite the origin of the internal thoracic artery. Trace it and the accompanying *vertebral vein* forward. The right vertebral vein normally enters the dorsal surface of the vena cava, and the left one, the brachiocephalic vein, but variation is common. The vertebral vessels soon enter the transverse foramina of the cervical vertebrae through which they continue, finally to enter the cranial cavity and help supply the brain.

A short *costocervical trunk* arises from the subclavian artery just distal to the origin of the vertebral and divides almost immediately into highest intercostal and deep cervical arteries. Sometimes these vessels arise independently from the subclavian. The *highest intercostal artery* extends caudally across the cranial ribs supplying those intercostal spaces drained by the highest intercostal vein. This vein, normally a tributary of the azygos, has been identified. The *deep cervical artery* extends dorsally to supply deep muscles of the neck. A major branch of it also passes cranial to the first rib and into the serratus ventralis muscle. The deep cervical artery is accompanied by the *deep cervical vein,* which usually drains into the vertebral vein shortly after this vein emerges from the transverse foramina. Occasionally the deep cervical vein enters the cranial vena cava independently.

The last branch of the subclavian artery is the *superficial cervical artery.* It extends deep to the subclavian vein and follows the *external jugular vein* cranially. Trace them both. The superficial cervical artery gives off one or more small branches that extend cranially, sometimes reaching the thyroid gland, but the main part of the artery continues laterally and dorsally to supply muscles on the craniolateral surface of the shoulder. A *superficial cervical vein,* a tributary of the external jugular, accompanies the distal part of the artery. One tributary of the superficial cervical vein, the *cephalic vein,* is often a conspicuous vessel, draining the lateral surface of the brachium.

After giving off these vessels, the subclavian artery and the satellite subclavian vein continue laterally into the armpit (axilla). These vessels change their names at this point to the *axillary artery* and *vein.* Major branches of the axillary artery are a *lateral thoracic artery* to the pectoral muscles and a *subscapular artery.* The latter passes between the subscapularis and supraspinatus muscles to supply deep shoulder muscles. Veins accompany the arteries but are usually not injected. When the axillary artery and vein enter the arm they are known as the *brachial artery* and *vein.*

Return to the brachiocephalic trunk and trace one of the *common carotid arteries* forward. It passes deep to the brachiocephalic vein and continues cranially, lying lateral to the trachea, supplying the trachea, thyroid gland, and other cervical structures before reaching the head. An *internal jugular vein,* which helps drain the inside of the skull, lies lateral to the common carotid through most of its course. The internal and external jugular veins unite slightly cranial to the subclavian vessels to form the *bijugular trunk* previously observed. Lymphatic vessels may also be seen entering the external jugular in this region. They are not ordinarily injected, but in any event they can be recognized by their unique appearance. Lymphatic vessels contain so many valves that they have alternating expanded and contracted portions and resemble a string of beads.

A cervical extension of the sympathetic cord and the vagus nerve can be found between the common carotid artery and the internal jugular vein. They are bound together by connective tissue to form a *vagosympathetic trunk,* but can easily be dissected apart. The vagus is the larger and passes superficial to the brachiocephalic trunk.

The major vessels in this region in human beings are essentially the same (Figs. 8–6 and 8–7). In addition to the azygos vein lying on the right side of the vertebral column, we have a hemiazygos that drains the left intercostal spaces before crossing the vertebral column to empty into the azygos. Our brachiocephalic trunk gives rise to only the right subclavian and right common carotid arteries; the left common carotid arises independently from the arch of the aorta, as does the left subclavian.

The major arteries described are derived from the embryonic aortic arches in the manner shown in Figure 8–8. All six of the primitive aortic arches appear during embryonic development and connect the ventral aorta (paired cranial to the fourth aortic arch) with the dorsal aorta (paired in the region of the aortic arches and for a short distance caudad). The first, second, and fifth aortic arches, the dorsal part of the right sixth aortic arch, the paired dorsal aortae between arches three and four, and the right paired dorsal aortae caudad to the entrance of the right subclavian (an intersegmental artery) disappear during development. The dorsal part of the left sixth arch persists during embryonic life as the *ductus arteriosus* which, as previously described, shunts blood from the pulmonary trunk directly to the dorsal aorta. It remains in the adult as the functionless *ligamentum arteriosum.*

The ventral portions of the sixth arches persist as the pulmonary arteries. The left fourth arch, together with part of the ventral and left dorsal aortae, forms the arch of the aorta. (Differential growth has the effect of shortening this arch and the adjacent dorsal aorta, so that the left subclavian of the adult leaves the arch of the adult aorta much closer to the common carotids than it does in the embryo.) The right fourth arch, plus a segment of the right dorsal aorta, forms the proximal part of the right subclavian. A splitting of the caudal part of the ventral aorta, and of the conus arteriosus, results in the direct origin of the arch of the aorta and the pulmonary trunk from the ventricles.

The paired ventral aortae between the fourth and third arches form the common carotids. The ventral aortae cranial to the third arch become the external carotids; the third arches, plus the dorsal aortae cranial to them, the internal carotids. The internal carotid of the embryonic mammal not only supplies the intracranial part of the head but also, by its stapedial branch passing through the stapes, much of the outside of the head. However, in the adults of most mammals the external carotid taps into the stapedial and pirates most, or all, of its peripheral distribution. If the external carotid takes it all over, the stapedial disappears.

(B) MAJOR VESSELS OF THE HEAD

Complete the skinning of the head, if this has not been done, and also remove most of the auricle on one side. Tributaries of the *external jugular vein* are superficial to other vessels in the head and must be considered first. About the level of the mandibular gland, the external jugular is formed by the confluence of the linguofacial and maxillary veins (Fig. 8–9). Trace the *linguofacial vein* forward. It soon receives on its medial side the *hyoid venous arch,* which crosses to the opposite side of the body, and, in turn, receives a deep vein from the larynx *(larengea impar).* At the caudoventral border of the mandible, the linguofacial is formed by the joining of a lingual and a facial vein. The *lingual vein* receives a small tributary from the lower lip and then extends deeply to drain the tongue and other structures in the floor of the mouth. It is accompanied by the hypoglossal nerve. It may have been cut in earlier dissections. The *facial vein* continues forward along the ventral border of the masseter muscle. Its major tributaries are a *deep facial vein* from beneath the ventral border of the masseter muscle, which has connections with venous plexuses in the orbit and on the palate, a *labial vein* from the upper lip, and a *vena angularis oculi* from the face in front of the eye.

Return to the level of the mandibular gland and trace the *maxillary vein* dorsally toward the parotid gland and base of the auricle. It is formed by the confluence of a *caudal auricular vein* from behind the ear and the top of the head, and a *superficial temporal vein* from in front of the ear. The superficial temporal receives tributaries from the ear, temporal muscle, and a deep branch connecting with the orbital and palatine venous plexuses. These plexuses, which will not be dissected, receive most of the drainage from inside the skull in the cat because the internal jugular veins are so small.

In order to trace the internal jugular vein and common carotid artery forward, reflect the mandibular gland, and the digastric and mylohyoid muscles. At the level of the caudal end of the larynx, the *common carotid artery* gives off a *cranial thyroid artery* and a muscular branch that crosses the longus coli muscle (Fig. 8–10). At the level of the cranial end of the larynx, the common carotid of most mammals splits into

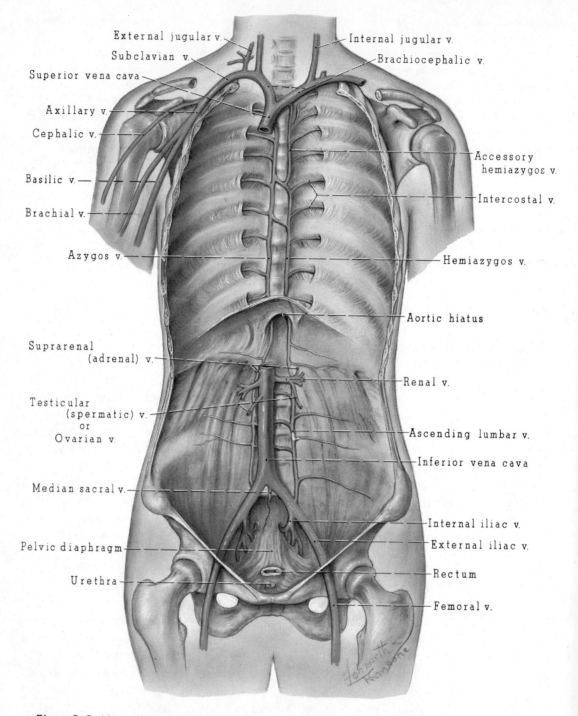

Figure 8–6 Ventral or anterior view of the major veins in a human being, exclusive of the hepatic portal system. (From Jacob and Francone: Structure and Function in Man. W.B. Saunders Company.)

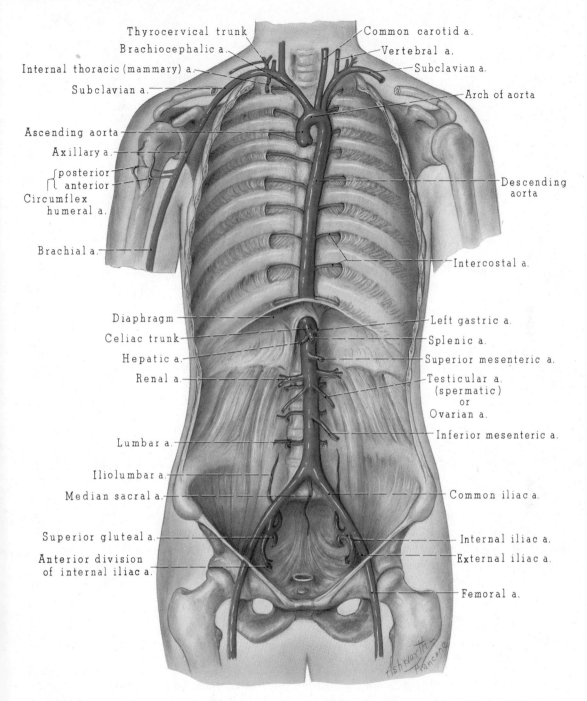

Thyrocervical trunk
Brachiocephalic a.
Internal thoracic (mammary) a.
Subclavian a.
Ascending aorta
Axillary a.
{posterior
{anterior
Circumflex
humeral a.
Brachial a.
Diaphragm
Celiac trunk
Hepatic a.
Renal a.
Lumbar a.
Iliolumbar a.
Median sacral a.
Superior gluteal a.
Anterior division
of internal iliac a.

Common carotid a.
Vertebral a.
Subclavian a.
Arch of aorta

Descending
aorta

Intercostal a.

Left gastric a.
Splenic a.
Superior mesenteric a.
Testicular a.
(spermatic)
or
Ovarian a.
Inferior mesenteric a.

Common iliac a.

Internal iliac a.
External iliac a.

Femoral a.

Figure 8–7 Ventral or anterior view of the major arteries in a human being. (From Jacob and Francone: Structure and Function in Man. W.B. Saunders Company.)

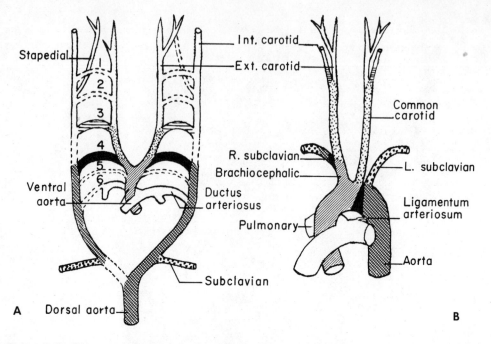

Figure 8–8 Diagrammatic ventral views of the mammalian aortic arches and their derivatives. *A*, embryonic condition; *B*, adult condition in a human being. (Slightly modified after Barry.)

internal and external carotids. An internal carotid is present in an embryonic cat, but it atrophies in the adult into an *internal carotid ligament* (Fig. 8–11). This ligament arises from the dorsal surface of the common carotid slightly caudal to the crossing of the hypoglossal nerve (Fig. 8–10). It then goes deep to the occipital artery, crosses the cranial cervical ganglion of the sympathetic cord, goes deep to the internal jugular vein, and enters the skull through the carotid canal.

After the origin of the internal carotid ligament, the common carotid is known as the *external carotid artery.* Three small branches arise just cranial to the ligament. A *laryngeal artery* goes to the cranial part of the larynx; an *occipital artery* extends dorsally, caudal to the tympanic bulla, and into the dorsal cervical muscles; and an *ascending pharyngeal artery* extends forward toward the bulla and into the skull. Ascending pharyngeal and occipital sometimes have a common origin from the external carotid.

Arterial blood reaches the brain of most vertebrates by way of the vertebral and internal carotid arteries, but the cat's internal carotid has lost this primitive function. To some extent, the ascending pharyngeal artery, which you have just seen, functionally replaces the internal carotid, but most of the arterial blood that would have gone to the brain via the internal carotid flows instead through the external carotid, then through a rete, or network, of fine arteries located deep in the orbit, and finally through an anastomotic artery to the circle of Willis and the brain (Fig. 8–11). The orbital venous sinus, which receives blood from the nasal area before draining into tributaries of the external jugular vein, is closely entwined with the arterial rete in such a way that the direction of venous blood flow is opposite that of the arterial flow. Venous blood returning from the nose is substantially cooler than the arterial blood, and its countercurrent juxtaposition to arterial blood in the rete reduces the temperature of arterial blood taking this route to the brain. We have here a mechanism that permits body temperatures to rise, yet prevents the brain temperature from increasing beyond a critical level.

Such mechanisms have been found in mammals that are subject to a considerable rise in body temperature from exposure to the sun (certain antelopes) or from intensive bursts of activity (cats), but that cannot reduce body temperature as a whole by losing much body water through panting or sweating, either because they live in a dry environment and must conserve water, or because they are heavily furred.

After the origin of the occipital and ascending pharyngeal, the external carotid gives rise to a number of branches supplying different parts of the outside of the head. These branches accompany the corresponding veins already observed. Lingual and facial arteries arise from the ventral surface of the external carotid (Fig. 8–10). The *lingual artery* enters the tongue. The *facial artery* follows the ventral border of the masseter muscle and supplies the jaws and facial structures.

Dorsal branches of the external carotid are (1) a *caudal auricular artery,* which extends dorsally behind the ear; (2) a *superficial temporal artery,* which extends dorsally in front of the ear; and (3) a *maxillary artery* that goes deep to the caudal border of the masseter to supply structures in the orbital and palatal regions.

Return to the *internal jugular vein.* It receives small tributaries from muscles at the base of the head, and then enters the skull through the jugular foramen to help drain the brain.

You have noticed that in the cat the external carotid artery supplies the entire outside of the head, and that certain of its branches enter the skull to help supply the brain. The vertebral artery also enters the skull. In human beings, the external carotid supplies primarily the outside of the head; the internal carotid and vertebral enter the skull and are the major arteries to the brain. The external jugular vein of the cat drains most of the head, both outside and inside the skull, although the vertebral vein, and to a small extent the internal jugular, help to drain the brain. These last two vessels are the main vessels draining the human brain; most of the drainage of our external jugular is limited to the outside of the head.

ARTERIES AND VEINS CAUDAL TO THE HEART

(A) VESSELS OF THE DORSAL THORACIC AND ABDOMINAL WALLS

After curving to the dorsal side of the body, the arch of the aorta is known as the *descending aorta.* Trace it caudally. As it passes through the thorax along the left side of the vertebral column, it gives off paired *intercostal arteries* to those intercostal spaces not supplied by the highest intercostals, small median branches to the esophagus, and also small branches to the bronchi, since the lungs, like the wall of the heart, need a separate arterial supply. The *thoracic portion* of the left *sympathetic cord* can be found at this time by carefully dissecting in the connective tissue near the heads of the ribs dorsal to the aorta. Enlargements along the cord are *sympathetic ganglia;* delicate strands passing dorsally are *communicating rami* to the spinal nerves. The left *vagus nerve* crosses the lateral surface of the arch of the aorta, passes dorsal to the root of the lung, and goes caudally along the esophagus. *Phrenic arteries* to the diaphragm may arise from the aorta before the aorta passes through the diaphragm, or from the

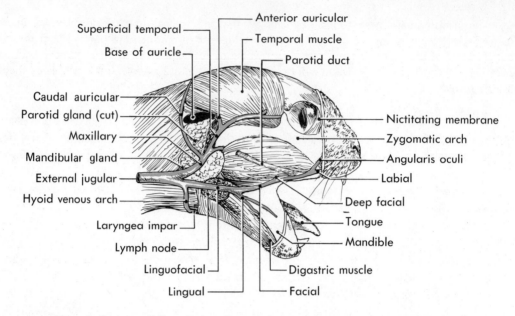

Figure 8–9 Lateral view of the tributaries of the external jugular vein of a cat.

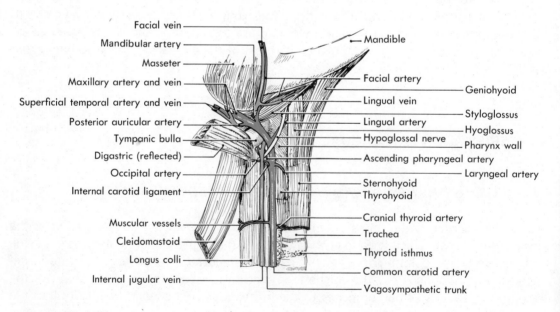

Figure 8–10 Lateroventral view of the internal jugular vein and carotid branches of a cat.

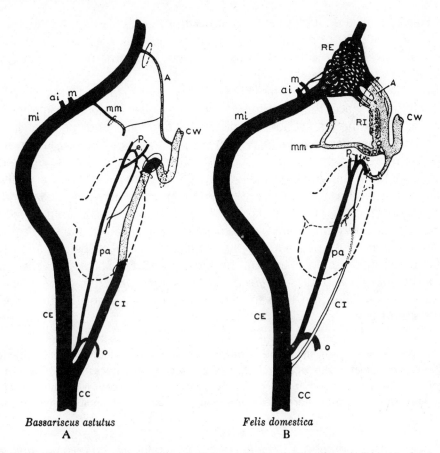

Figure 8–11 Diagrams in ventral view of the carotid circulation of a primitive carnivore, a member of the raccoon family (*A*) and the domestic cat (*B*). Stippled parts are intracranial, or run through canals. The position of the tympanic bulla is shown by broken lines. Note the vestigial nature of the proximal part of the internal carotid of the cat (shown in outline), and the way in which other vessels have enlarged to carry blood to the brain. Some other carnivores have an intermediate condition. Abbreviations: *A,* anastomotic artery; *ai,* inferior alveolar; *CC,* common carotid; *CE,* external carotid; *CI,* internal carotid; *CW,* circle of Willis located on ventral surface of brain; *e,* eustachian; *m,* masseteric; *mi,* internal maxillary; *mm,* median meningeal; *o,* occipital; *p,* pharyngeal; *pa,* ascending pharyngeal; *RE,* external rete; *RI,* internal rete. (From Davis and Story: The carotid circulation in the domestic cat. Zoological Series, Field Museum of Natural History, Vol. 28.)

last intercostals, or they may arise from vessels caudal to the diaphragm (first lumbar, cranial abdominal, celiac).

The *caudal vena cava,* or postcaval, was seen entering the heart. Trace it caudad. As it passes through the diaphragm, it receives several small *phrenic veins* and then disappears in the liver. Scrape away tissue from the cranial surface of the right medial lobe of the liver and find the entrance of several large *hepatic veins.* The major part of the caudal vena cava, however, passes through the right lateral and caudate lobes; it should also be exposed by scraping away liver tissue. Other hepatics, most very small, will be seen entering.

Push the abdominal viscera to the right and find the aorta emerging from the diaphragm. Just after emerging, it gives rise to two ventral vessels—first a *celiac trunk* and then a *cranial mesenteric artery* (Fig. 8–12) which supply most of the abdominal viscera. Trace them later. *Celiac* and *mesenteric ganglia* of the sympathetic nervous system lie at the base of the cranial mesenteric artery. They receive one or more *splanchnic nerves* from the sympathetic cord and send out minute branches which travel along the vessels to the viscera.

Slightly caudal to the cranial mesenteric artery, the aorta lies to the left side of the caudal vena cava. Trace these two vessels to the pelvic region. Their most cranial, paired branches are the cranial abdominal arteries and veins and the renal arteries and veins. The *renal arteries* and *veins,* which supply the kidneys, are larger and more obvious. Those of the right side of the body lie slightly cranial to those of the left side, since the right kidney is more cranially situated than the left one. Carefully dissect away fat from around each kidney so that you can lift up the lateral edge and look at the muscles dorsal to it. The vessels you see supplying the abdominal wall are the *cranial abdominal artery* and *vein.* Trace them toward the aorta and vena cava. Before joining these vessels, they pass and supply a small, hard, oval-shaped nodule embedded in the fat between the cranial end of the kidney and the aorta and vena cava. This nodule is the *suprarenal* (adrenal) *gland* — one of the endocrine glands. The cranial abdominal vessels usually join the vena cava and aorta just cranial to the renal vessels, but they sometimes join the renals.

The suprarenal gland is an endocrine gland of dual origin. Its medullary portion, derived from postganglionic sympathetic cells of neural crest origin, secretes hormones which assist sympathetic stimulation in adjusting the body to meet conditions of stress. Its cortical portion, of mesodermal origin, secretes numerous steroid hormones involved in many aspects of metabolism and also in sexual differentiation.

The next paired branches of the aorta are the small *testicular* or *ovarian arteries,* depending on the sex. They pass to the gonads accompanied by the *testicular* or *ovarian veins.* The ovaries are small, oval bodies lying near the cranial ends of the Y-shaped uterus. The testes have descended into the scrotum, and, in doing so, each has made an apparent hole (the *inguinal canal*) through the body wall in the region of the groin. The testicular vessels and the sperm duct *(ductus deferens)* can be seen passing through these canals (Fig. 9–4, p. 176). The right gonadial vein enters the caudal vena cava; the left one may too, but it normally enters the left renal.

A *caudal mesenteric artery* to the colon leaves the ventral surface of the aorta caudal to the gonadial arteries. Trace it later. Caudal to this vessel, the aorta gives rise to a pair of *deep circumflex iliac arteries* (iliolumbar arteries), which pass laterally to the musculature and body wall lying ventral to the ilia. Satellite *deep circumflex iliac veins* accompany the arteries and enter the caudal vena cava. The rest of the lumbar musculature is supplied by several *lumbar arteries* and *veins* which can be found by dissecting along the dorsal surface of the aorta and vena cava between the renal and deep circumflex vessels. The lumbars are single vessels where they attach to the aorta and vena cava but they bifurcate distally. Caudal to the deep circumflex vessels, the aorta and caudal vena cava give rise to the iliac vessels supplying the pelvic region and leg. Trace them later.

The human suprarenal glands are triangular-shaped organs that cap the cranial ends of the kidneys, but the blood vessels on the dorsal thoracic and abdominal walls are essentially the same as in the cat (Figs. 8–6 and 8–7). We often have several vessels to the suprarenal glands; the lumbar vessels are paired, and the deep circumflex iliac vessels join the external iliacs.

(B) VESSELS OF THE ABDOMINAL VISCERA

Blood reaches the abdominal parts of the digestive tract and the spleen through branches of the celiac, cranial mesenteric, and caudal mesenteric arteries. The veins draining these organs form the *hepatic portal system,* for they all drain into a portal vein which carries the blood to capillary-like spaces (sinusoids) in the liver rather than directly to the caudal vena cava. As the blood passes through the liver it comes into intimate contact with the hepatic cells. Excess food products in the blood coming from the digestive tract after a meal are stored in the hepatic cells largely in the form of glycogen, and deficiencies in the food content of the blood between meals is made up from food stored in the cells. Numerous other metabolic conversions also occur here. Liver sinusoids are drained by the hepatic veins which you have seen entering the caudal vena cava.

Return to the celiac trunk and mesenteric arteries where they leave the aorta. Remove surrounding connective tissue and the sympathetic ganglia and trace the *celiac trunk* a short distance until it divides into three branches (Fig. 8–13)—a *lienic artery* to the spleen, a *left gastric artery* to the lesser curvature of the stomach, and a *hepatic artery* to the liver, pancreas, duodenum, and part of the stomach. More distal parts of these vessels will be seen with the veins. The distribution of the *cranial mesenteric artery* to most of the small intestine and adjacent parts of the colon can be seen by stretching the mesentery. The *caudal mesenteric artery* supplies the descending colon and rectum (Fig. 8–12).

Although not injected, the *portal vein* can be found in the lesser omentum where it lies dorsal to the bile duct and forms the ventral border of the epiploic foramen. Trace it caudad (Fig. 8–13). As it passes dorsal to the pylorus it receives a small, and often inconspicuous, *right gastric vein* from the pyloric region of the stomach, and a larger *gastroduodenal vein.* The latter is formed by the confluence of a *cranial pancreatico-duodenal vein* draining much of the duodenum and pancreas and a *right gastroepiploic vein* from the greater curvature of the stomach and greater omentum. *Cranial pancreaticoduodenal, right gastroepiploic, gastroduodenal,* and *gastric arteries* accompany the veins. All are derived from the hepatic artery which can be seen on the left side of the epiploic foramen. After giving rise to these arteries, the hepatic artery follows the portal vein to the liver.

Push the stomach forward and tear through the part of the greater omentum going to the spleen and dorsal body wall. Carefully dissect away the tail of the pancreas, which extends toward the spleen, and notice that the portal vein is formed by the confluence of two tributaries—a lienogastric vein entering from the left side of the animal, and a much larger cranial mesenteric vein. Trace the *lienogastric vein* by continuing to dissect away pancreatic tissue. Its tributaries are a *left gastric vein,* which accompanies the left gastric artery and drains the lesser curvature of the stomach, and a *lienic vein,* which accompanies the lienic artery to the spleen. A *left gastroepiploic artery* and *vein* can be found on the greater curvature of the stomach. They join the lienic vessels at several points.

Now trace the *cranial mesenteric vein.* One of its tributaries is the *caudal mesenteric vein* from the large intestine. Parts of this vein are accompanied by branches of the cranial mesenteric artery, and part by branches of the caudal mesenteric artery. Other tributaries of the cranial mesenteric vein accompany branches of the cranial mesenteric artery to the caudal parts of the pancreas and duodenum *(caudal pancreaticoduodenal vein* and *artery),* and to the numerous coils of the small intestine *(jejunoiliac veins* and *arteries).*

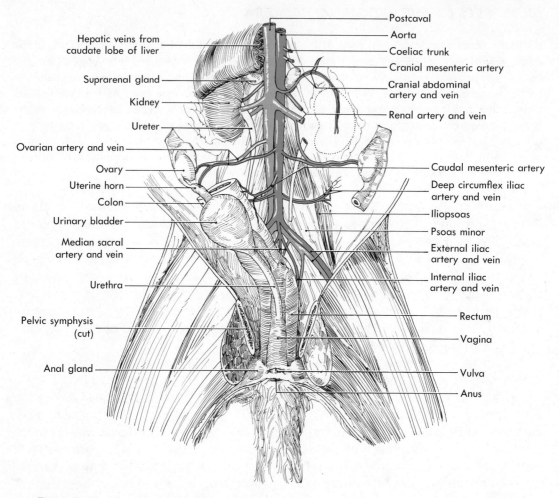

Figure 8–12 Ventral view of the abdominal portions of the caudal vena cava of a female cat. The pelvic canal has been cut open, and the left kidney and uterine horn have been omitted to show deep vessels.

The human digestive organs and spleen are supplied and drained by vessels similar to those of the cat. The pattern of the arteries is substantially the same, but there are a few differences in the hepatic portal system. The human left gastric vein enters the portal vein independently, and the inferior (caudal) mesenteric vein joins the lienic rather than the superior (cranial) mesenteric vein.

(C) VESSELS OF THE PELVIC REGION AND HIND LEG

Return to the caudal ends of the caudal vena cava and aorta. The terminal branches of the aorta pass superficial to the tributaries of the vena cava as they enter the pelvis. In order to see the pelvic vessels clearly, open the pelvic canal. This is a simple procedure in the female. Cut the ventral ligament of the bladder and push it away from the cranioventral border of the pelvic girdle. Then take a scalpel, cut through the muscles on the ventral face of the girdle, and continue to cut right through the midventral symphysis. Bone scissors may be used, but this is not usually necessary if you keep in the midventral line. Now take a firm grip on the thighs and bend them as far dorsally as you can. The procedure for the male is the same, but one must use more caution to avoid reproductive ducts. First locate the narrow cremasteric pouches that extend from the inguinal canals, across the ventral surface of the girdle and into the skin of the scrotum (Fig. 9–4, p. 176). They should be pushed aside before cutting. Also locate the penis emerging from the caudal end of the pelvic canal; avoid cutting it. After the canal is opened, carefully pick away fat and connective tissue from around the vessels, bladder, and rectum. Insofar as possible, confine your dissection to one side and do not injure parts of the urogenital system.

An *external iliac artery* extends from the aorta laterally and caudally toward the body wall and leg. It is accompanied distally by the *external iliac vein* (Fig. 8–14). An *internal iliac artery* and *vein* enter the pelvic cavity. The iliac arteries arise independently from the aorta in the cat, but the external and internal iliac veins unite to form a *common iliac vein* before entering the caudal vena cava.

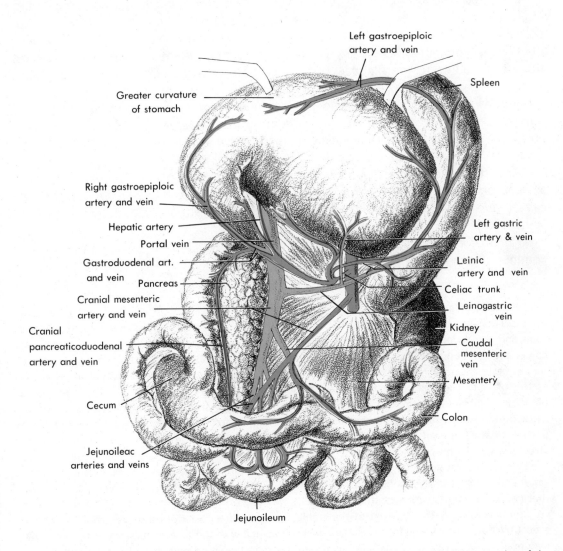

Figure 8-13 Ventral view of the hepatic portal system of veins and accompanying arteries of the cat. The stomach has been pulled forward and the tail of the pancreas dissected away.

Trace the external iliac vessels. Usually just inside the abdominal wall, the external iliac artery and vein give off from their caudomedial surface a *deep femoral artery* and *vein,* which extend deep into the thigh. A *caudal epigastric artery* and *vein* can be seen on the peritoneal surface of the rectus abdominis. They anastomose cranially with the cranial epigastric vessels previously seen. The caudal epigastric artery is usually a branch of the deep femoral, but it may arise directly from the external iliac near the deep femoral. An *external pudendal artery* and *vein* can be found in the mass of fat in the region of the groin. They continue through the fat and supply the external genitalia. The artery may be a branch of the caudal epigastric or of the deep femoral. The caudal epigastric and external pudendal veins normally form a short, common *pudendoepigastric trunk* before they join the deep femoral vein. After giving rise to these vessels, the external iliac vessels perforate the abdominal wall and enter the leg as the *femoral artery* and *vein.* Additional, major branches of these vessels are shown in Figure 8–14.

Now trace the internal iliac vessels. Near its origin from the aorta, the internal iliac artery gives rise to a *vesical artery* to the urinary bladder. The artery is a remnant of the large umbilical artery of the embryo which goes to the placenta. The proximal part of the embryonic umbilical artery persists as a part of the internal iliac artery and as the vessel leading to the bladder, but the portion from the bladder to the umbilicus atrophies. Deeper within the pelvic cavity, the internal iliac artery gives rise to one or two *gluteal arteries* to deep pelvic muscles and to an *internal pudendal artery* to remaining pelvic viscera. *Gluteal* and *internal pudendal veins* accompany the arteries and drain into the internal iliac vein. A small *vesical vein* normally joins the internal pudendal.

After the iliac arteries have branched off, the aorta continues caudad as a very small vessel across the sacrum *(median sacral artery)* and into the tail *(caudal artery).* A *caudal vein* leads to a *median sacral vein* which enters either, or both, common iliac veins.

The major difference in the human vessels in the region is the presence of a common iliac artery from which the external and internal iliac arteries arise (Fig. 8–7).

Parts of the mammalian venous system resemble parts of the system in other vertebrates, but some parts have changed considerably. The major change is the conversion of parts of the hepatic veins and the primitive cardinal system into a caval and an azygos system. The way in which this comes about is best understood by recourse to the embryonic development of the veins in a mammal.

An early mammal embryo (Fig. 8–15, *A*) has a cardinal and an incipient renal portal system (a system of veins carrying blood from the hind legs into capillaries in the kidneys), for some of the blood in the caudal part of the posterior cardinals passes through the kidneys to a pair of subcardinals. In this stage the mammalian embryo is similar to a fish, except that in an adult fish the portion of the posterior cardinals situated just caudal to the cranial attachment of the subcardinals atrophies, and the flow of blood through the kidneys and into the subcardinals is mandatory.

Later in the development (Fig. 8–15, *B*), the right hepatic enlarges, and a caudal extension of the vessel unites with the right subcardinal to form the proximal part of the caudal vena cava. The two subcardinals also unite with each other.

Still later (Fig. 8–15, *C*), most of the cranial portion of the posterior cardinals atrophies, but the caudal portion on each side forms a large vessel connecting with the subcardinals. The essentially new feature of the trunk veins of mammals is the subsequent formation of a pair of *supracardinals* (Fig. 8–15, *C*), connecting cranially and caudally with the remnants of the posterior cardinals. The supracardinals also become connected with the subcardinals by a pair of *subsupracardinal anastomoses,* or the *renal collar.* This connection makes possible the elimination of most of the caudal portion of the posterior cardinals (the renal portal system of lower vertebrates).

During subsequent development, the supracardinals become divided into a cranial thoracic portion and a caudal lumbar portion (Fig. 8–15, *D*). The right subsupracardinal anastomosis and lumbar portion of the supracardinal enlarge, while those of the left side do not. Renal veins grow out from the renal collar to the definitive kidneys, which have migrated cranially.

By the adult stage (Fig. 8–15, *E*), all but the most caudal segments of the posterior cardinals are lost, the left subsupracardinal anastomosis is lost, and the caudal vena cava is extended caudad by the enlargment of the right subsupracardinal anastomosis and lumbar portions of the supracardinals. In some mammals, only the right supracardinal is involved, but in the cat the right enlarges and absorbs the lumbar portion of the left supracardinal. Thus, the adult caudal vena cava is formed of the right hepatic, a caudal outgrowth from the right hepatic, the middle section of the right subcardinal, the right subsupracardinal anastomosis, the lumbar portion of the supracardinals (especially the right supracardinal), and a small segment of the caudal end of the posterior cardinals. The renal veins are formed primarily by outgrowths from the renal collar, but the left subcardinal contributes to the left renal vein. The genital veins are formed from the subcardinals plus a small segment of the posterior cardinals; the cranial abdominals are formed from the subcardinals.

While these changes are taking place, the thoracic portion of the left supracardinal disappears. But the thoracic portion of the right supracardinal, together with the proximal end of the right posterior cardinal, forms the azygos. The cat does not have a left highest intercostal, but the vessel develops in such mammals as have it (rabbit) from the stump of the left posterior cardinal.

The formation of the cranial vena cava is a simpler affair. In a mammal such as a rabbit, the condition shown in Figure 8–15, *C* persists. The two cranial venae cavae represent the common cardinals plus the proximal portion of the anterior cardinals. The more distal portion of the anterior cardinals is represented by the internal jugular. The external jugular is a new outgrowth. But in mammals such as the cat and human being, a cross anastomosis, which is to be the left brachiocephalic, develops between the anterior cardinals (Fig. 8–15, *D*). The right cranial vena cava is formed as above, but a left one does not form, for the proximal portion of the left anterior cardinal atrophies. The left common cardinal persists, however, as the coronary sinus.

BRONCHI AND INTERNAL STRUCTURE OF THE HEART

Cut the great vessels near the heart of your specimen, remove the heart, and examine the roots of the lungs. The bifurcation of the trachea into bronchi referred to earlier (p. 135) can now be exposed. Trace a bronchus into a lung and notice that it subdivides repeatedly into smaller and smaller passages that terminate in clusters of thin-walled, microscopic sacs (the *alveoli*) where gas exchange occurs. This entire complex of passages is called the *respiratory tree.*

Again identify the chambers of the heart and the great vessels entering and leaving it as they appear in a ventral view (Fig. 8–16, *A*). Carefully clean the dorsal surface of the heart and identify the chambers and vessels in this view (Fig. 8–16, *B*). Internal features can be seen by dissecting either the heart of your own specimen or a separate sheep heart. The latter is preferable, if material is available, for the structures are larger and the chambers are not clogged with the injection mass. If a sheep heart is used, you will have to remove the pericardial sac, and clean and identify the great vessels. They are similar to those of the cat, except that both of the subclavian and common carotid arteries leave the arch of the aorta by a common brachiocephalic. A small left cranial vena cava is also present in the sheep, and the ligamentum arteriosum is conspicuous.

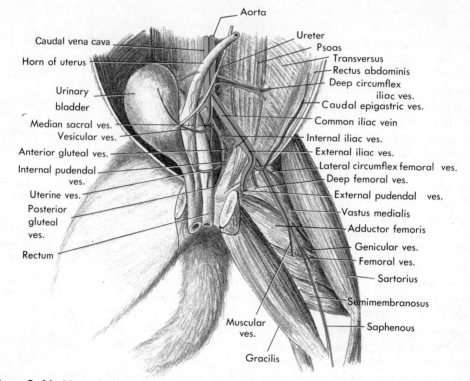

Aorta

Caudal vena cava

Horn of uterus

Urinary
bladder

Median sacral ves.
Vesicular ves.

Anterior gluteal ves.

Internal pudendal
ves.

Uterine ves.

Posterior
gluteal
ves.

Rectum

Ureter
Psoas
Transversus
Rectus abdominis
Deep circumflex
iliac ves.
Caudal epigastric ves.

Common iliac vein

Internal iliac ves.
External iliac ves.
Lateral circumflex femoral ves.
Deep femoral ves.
External pudendal ves.

Vastus medialis
Adductor femoris

Genicular ves.
Femoral ves.

Sartorius

Semimembranosus
Saphenous

Muscular
ves.

Gracilis

Figure 8–14 Ventral view of the distribution of the left external and internal iliac artery and vein in a female cat. The pelvic canal has been opened and the pelvic viscera pushed to the specimen's right side.

Open the right atrium by making an incision that extends from the auricle into the caudal vena cava; the left atrium, by an incision extending from its auricle through one of the pulmonary veins (Fig. 8–16). To open the ventricles, first cut off the apex of the heart in the transverse plane. Cut off a sufficient amount to expose the cavities of both ventricles. Then make a cut through the ventral wall of the right ventricle and extend it from the cut surface made by removing the apex into the pulmonary artery. Open the left ventricle by making an incision through its ventral wall that extends from the cut surface as far forward as the base of the arch of the aorta. Clean out the chambers of the heart if necessary.

Find the entrances of the *right cranial vena cava* and *caudal vena cava* into the *right atrium.* The entrance of the *coronary sinus* (cat), or *left cranial vena cava* (sheep), lies just caudal to the entrance of the caudal vena cava. The extent of the coronary sinus can be determined by probing. Also find the entrances of the *pulmonary veins* into the *left atrium.* The atria have relatively thin muscular walls; however, the muscles in their auricles form prominent bands known as *pectinate muscles* because they resemble a comb, or pecten.

The two atria are separated by an *interatrial septum.* Examine the septum from the right atrium, and you will find an oval-shaped depression, the *fossa ovalis,* beside the point at which the caudal vena cava enters. Put your thumb in one atrium and forefinger in the other, and palpate this region. You will feel that the septum is

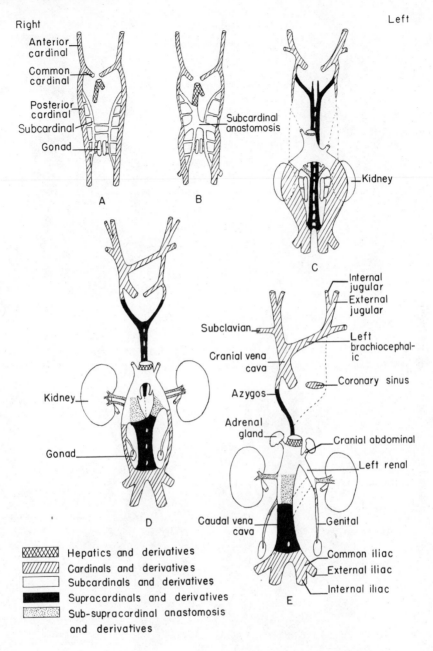

Right

Anterior cardinal

Common cardinal

Posterior cardinal

Subcardinal

Gonad

Left

Subcardinal anastomosis

Kidney

A

B

C

Kidney

Gonad

D

Internal jugular

External jugular

Subclavian

Cranial vena cava

Azygos

Adrenal gland

Caudal vena cava

Left brachiocephalic

Coronary sinus

Cranial abdominal

Left renal

Genital

Common iliac

External iliac

Internal iliac

E

Hepatics and derivatives

Cardinals and derivatives

Subcardinals and derivatives

Supracardinals and derivatives

Sub-supracardinal anastomosis and derivatives

Figure 8–15 A series of diagrams ranging from a young embryo, *A*, to an adult, *E*, to show the development of the major veins of the cat from the primitive cardinal and renal portal system. All are ventral views. For explanation, see text. (Slightly modified after Huntington and McClure: The development of the veins in the domestic cat. Anatomical Record, Vol. 20.)

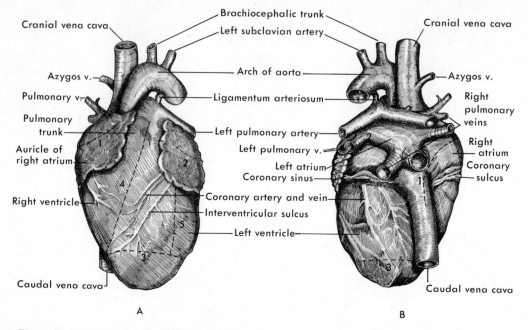

Figure 8–16 Ventral (*A*) and dorsal (*B*) views of the heart and great vessels of the cat. The incisions to be made to open the heart are shown by dashed lines.

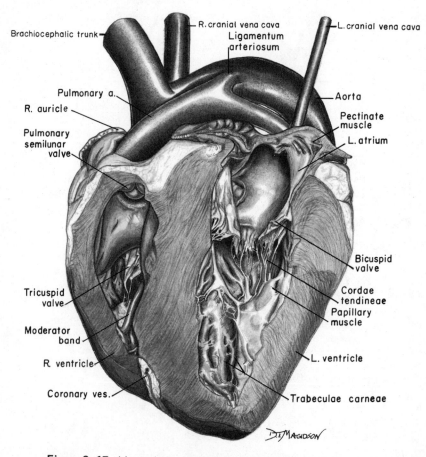

Figure 8–17 Ventral view of a dissection of the sheep heart.

unusually thin here. During embryonic life there is a valved opening, the *foramen ovale,* through the septum at this point, and much of the blood in the right atrium (mostly blood coming in by the caudal vena cava) is sent directly to the left atrium and out to the body. This opening closes at birth.

The *atrioventricular openings* will be seen in the floor of the atria. The right one is guarded by the *right atrioventricular,* or tricuspid, *valve* which consists of three flaps; the left one by the *left atrioventricular,* or bicuspid, *valve,* which consists of two flaps (Fig. 8–17). Since these flaps extend into the ventricles, they can be seen better from that aspect. Note that little tendinous cords *(chordae tendineae)* connect the margins of the flaps with the walls of the ventricles. Many of the chordae attach onto papilla-like extensions of the ventricular muscles *(papillary muscles).* The chordae tendineae may help to open the valves, but in any case they prevent the valves from everting into the atria during ventricular contraction.

Notice that the ventricles are separated from each other by an *interventricular septum* and that the walls of the ventricles are much thicker than those of the atria. The left ventricular wall is also much thicker than the right one since it pumps blood a greater distance. In addition to the papillary muscles, the inside of the ventricular walls bears irregular bands *(trabeculae carneae),* and sometimes bands that cross the lumen *(moderator bands).* There is a particularly prominent moderator band in the right ventricle of the sheep. Moderator bands are believed to prevent the overdistention of the ventricle.

Notice where the pulmonary trunk and arch of the aorta leave the ventricles. Three pocket-shaped, *semilunar valves* are located in the base of each vessel. Those in the pulmonary artery are known as the *pulmonary valve;* those in the aorta, as the *aortic valve.* The two coronary arteries leave from behind two of the semilunar valves in the aorta. One has probably been cut through.

The human heart lacks moderator bands, but in other respects is the same as the sheep heart.

LYMPHATIC SYSTEM

The relation of the lymphatic to the cardiovascular system was considered in the introduction to this chapter (p. 141). Although the lymphatic system is not conspicuous enough in most vertebrates to be studied easily, parts, at least, of the system can be seen in mammals even though it has not been specially injected.

The major lymphatic vessel of the body is the *thoracic duct* (Fig. 8–18). This is a brownish vessel that can be found in the left pleural cavity just dorsal to the aorta. Sometimes the vessel is divided into two or more channels. Trace it forward. It passes deep to most of the arteries and veins at the base of the neck, and then curves around to enter the left external jugular beside the entrance of the internal jugular (Fig. 8–4). Now trace it caudad. It passes through the diaphragm dorsal to the aorta and, dorsal to the origin of the celiac and cranial mesenteric arteries, expands into a sac called the *cisterna chyli.*

Next stretch out a section of the mesentery supporting the small intestine and hold it up to the light. Very small lymphatic vessels, in this case called *lacteals* because absorbed fat passes through them, can be seen outlined by little streaks of fat. These ultimately lead into an aggregation of *mesenteric lymph nodes* (the pancreas of Aselli)

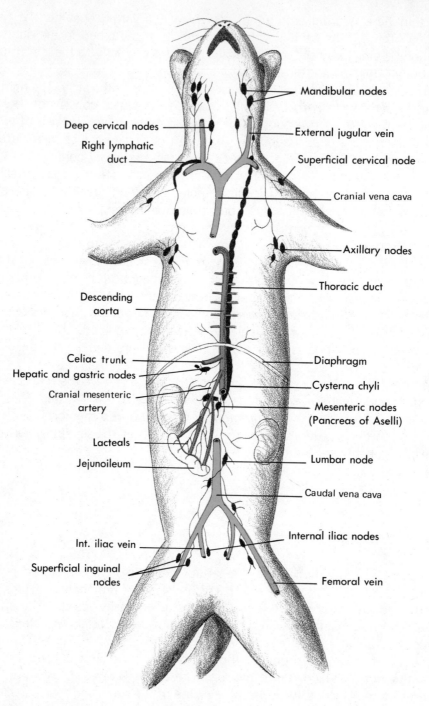

Figure 8–18 Diagrammatic ventral view of the major lymphatic vessels and groups of nodes in a cat.

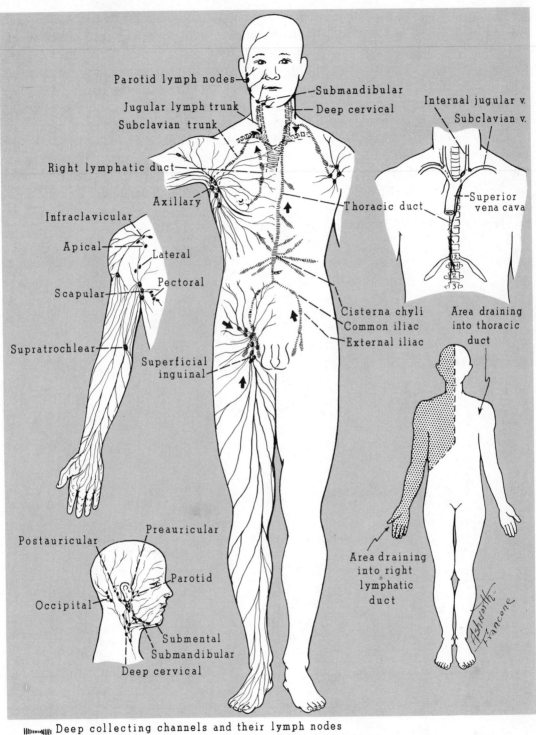

Parotid lymph nodes

Jugular lymph trunk

Subclavian trunk

Right lymphatic duct

Axillary

Infraclavicular

Apical

Lateral

Scapular

Pectoral

Supratrochlear

Superficial inguinal

Submandibular

Deep cervical

Internal jugular v.

Subclavian v.

Thoracic duct

Superior vena cava

Cisterna chyli

Common iliac

External iliac

Area draining into thoracic duct

Area draining into right lymphatic duct

Postauricular

Preauricular

Parotid

Occipital

Submental

Submandibular

Deep cervical

▥▥▥ Deep collecting channels and their lymph nodes

━━ Superficial collecting channels and their lymph nodes

Figure 8–19 The lymphatic system of a human being. (From Jacob and Francone: Structure and Function in Man. W.B. Saunders Company.)

located at the base of the mesentery. The mesenteric lymph nodes, in turn, are drained by one or more larger lymphatics that pass along the cranial mesenteric artery to the cisterna chyli. These vessels have probably been destroyed. Lymphatic vessels from the stomach, liver, pelvic canal, and hind legs also pass to the cisterna chyli, but they are hard to see. Thus, the cisterna chyli receives all the lymphatic drainage of the body caudal to the diaphragm and passes it on to the thoracic duct. The thoracic duct receives the lymphatic drainage of the thorax as it ascends through this region.

Other lymphatic vessels, which parallel the larger veins, drain the arms, neck, and head. Those of the left side enter the thoracic duct or the left external jugular close to the entrance of the thoracic duct. Those of the right side enter the right external jugular near the entrance of the internal jugular, either independently or by a short common trunk (the *right lymphatic duct*).

The pattern in human beings is substantially the same (Fig. 8–19).

Chapter Nine

THE EXCRETORY AND REPRODUCTIVE SYSTEMS

The excretory system plays an important role in eliminating the nitrogenous waste products of cellular metabolism and helps to control the water and salt balances of the body. The reproductive system has an entirely different function, namely, perpetuating the species. However, the two must be considered together morphologically, for in the males of most vertebrates excretory passages are utilized for the transport of the sperm, and in some cases the female genital ducts develop from excretory ducts. In view of this intimate morphological association, the two systems are sometimes referred to as the urogenital system.

HISTORY OF THE KIDNEY

A brief consideration of the history of the kidneys and their ducts is a prerequisite to an understanding of the urogenital system. The functional units of the kidneys are the *renal tubules* (nephrons). In all vertebrates they develop embryonically from a pair of bands of nephrogenic tissue *(nephric ridges)* located dorsal to the coelom between the somites and lateral plate mesoderm. In the ontogeny of an amniote, a pronephric kidney *(pronephros)* is succeeded by a *mesonephros* which, in turn, is succeeded by the definitive *metanephros* (Fig. 9–1). These kidneys have a linear relationship from cranial to caudal along the nephric ridge. The pronephros forms the *archinephric duct,* which extends caudad to the cloaca. The mesonephric tubules tap into this duct (then sometimes called the *mesonephric* or wolffian *duct,* but the metanephros is drained by a *ureter* which develops as a craniad outgrowth from the caudal end of the archinephric duct.

During the embryonic development of an amniote, there is a gradual caudad differentiation of tubules along the nephric ridge. As caudal tubules differentiate and begin to function, the cranial ones lose their excretory function, and many atrophy. The loss of a urinary function by the more cranial tubules also occurs during the evolution of vertebrates. In the adults of very primitive fishes, functional nephrons, one pair per body segment, occupy nearly the entire length of the nephric ridge. During subsequent phylogeny, there is a multiplication of tubules caudally, a complete loss of pronephric tubules in the adult, and a take-over of the more cranial mesonephric tubules by the male genital system. Reptiles resemble mammals in having a metanephros, whereas amphibians and most fishes have a condition intermediate between a metanephros and the primitive fish condition.

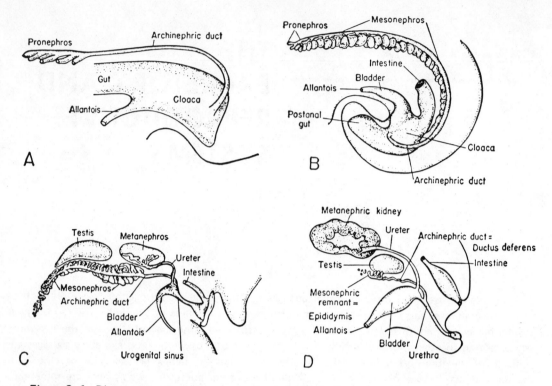

Figure 9–1 Diagrams to show the succession of kidneys in an amniote embryo. All are lateral views as seen from the left side. The genital details in these diagrams are those of a male. *A*, an early stage in which the pronephros and its duct are formed; *B*, mesonephric tubules tapping into the archinephric duct; *C*, pronephros degenerating, mesonephros functional, metanephros and ureter forming; *D*, definite stage in which the metanephros is functional and the remnants of the mesonephros and its duct are taken over by the male genital system. (From Romer: The Vertebrate Body.)

GENITAL DUCTS AND THEIR RELATION TO EXCRETORY DUCTS

In very primitive fishes, such as the cyclostomes, the gametes of both sexes are discharged into the coelom and pass to the outside through a pair of genital pores. This may have been the ancestral condition, but in other vertebrates ducts have evolved that transmit the gametes. In the embryo of either sex, primordia for the ducts of both the male and female are laid down during a *sexually indifferent stage.* An *oviduct* develops either from a splitting of the archinephric duct, or from a folding of the coelomic epithelium. In addition to the primordium of an oviduct, the same embryo acquires a series of cords, the *cords of the urogenital union (rete testis),* that connect the gonad with the cranial mesonephric tubules, and through them with the archinephric duct. Thus, two potential routes are present for gamete transport (Figs. 9–2, *C*).

During the subsequent differentiation of the female, the oviducts develop into the uterine tubes, the uterus, and part of the vagina, while the cords of the urogenital union and the adjacent parts of the mesonephros degenerate. In adult female amniotes, which have a metanephros and ureter, the mesonephros is represented by minute, functionless groups of tubules called the *epoophoron* (more cranial mesonephric tubules) and *paroophoron* (more caudal tubules), and the archinephric duct degenerates or forms a vestige known as the *longitudinal duct of the epoophoron* (Fig. 9–2, *E*).

During the subsequent differentiation of the male (Fig. 9–2, *D*), the oviduct degenerates (although some vestiges may persist), and the route through the kidney becomes the functional pathway for the sperm. In adult amniotes, the testis and the cranial end of what was the mesonephros are close

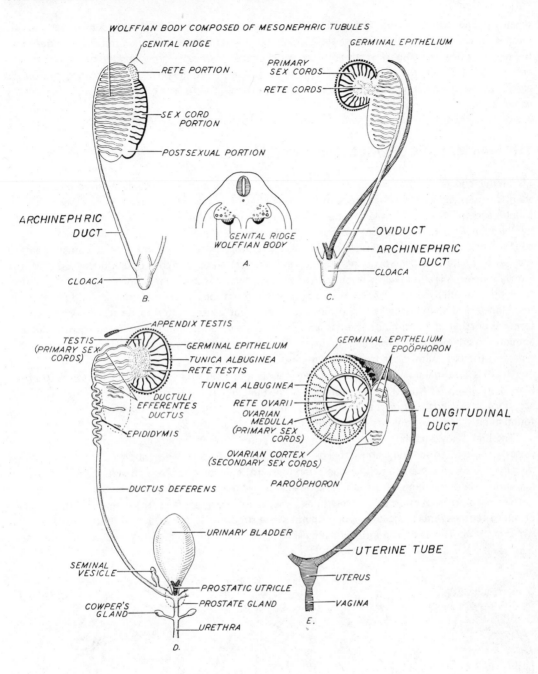

Figure 9–2 Diagrams of the development of the mammal genital system. *A*, cross section of an early embryo showing the location of the mesonephros (wolffian body) and developing gonad (genital ridge); *B*, ventral view of an early embryo; *C*, the indifferent stage; *D*, differentiation of the male condition. The primary sex cords of the gonad of the indifferent stage become the seminiferous tubules. *E*, Differentiation of the female conditions. The primary sex cords regress, and the follicles develop from secondary sex cords. (From Turner: General Endocrinology.)

together. The cords of the urogenital union constitute the *rete testis;* the cranial part of the mesonephros constitutes the *head of the epididymis,* and its tubules are called the *ductuli efferentes;* the highly coiled cranial part of the archinephric duct constitutes the *body* and *tail of the epididymis* and is called the *ductus epididymidis:* and the rest of the archinephric duct is called the *ductus deferens.* The more caudal parts of the mesonephros sometimes form a functionless vestige called the *paradidymis.*

THE MAMMALIAN UROGENITAL SYSTEM

In the course of their evolution through reptiles to mammals, vertebrates have become well adapted to an active, terrestrial mode of life. There has been a large increase in the number of kidney tubules and this makes possible the elimination of the large volume of nitrogenous wastes produced by the high level of metabolism of mammals. Whereas a kidney of a salamander such as *Necturus* contains approximately 50 tubules in its caudal excretory portion, the number in a small mammal such as a mouse is on the order of 20,000, and we have an estimated 1,000,000 to 4,000,000 per kidney. Nitrogen is eliminated without an excess loss of body water. In mammals a special segment of the kidney tubule, the *loop of Henle,* concentrates sodium ions in the intercellular fluid deep within the kidney and thereby makes an osmotic gradient that causes water to be reabsorbed from the terminal portion of the tubule. There is a more complete separation of excretory and genital ducts in amniotes than in anamniotes, and there is also a division of the cloaca in most mammals that separates the urogenital tract from the digestive tract.

Reptiles and mammals can also reproduce upon the land and do not need to return to water. Mating on land is facilitated by the evolution in the male of a *copulatory organ* and the elaboration of *accessory sex glands* that secrete the seminal fluid in which the sperm are carried. These organs are found in only a few anamniotes in which internal fertilization occurs.

The high body temperature of most mammals poses one reproductive problem because the final stages of sperm formation cannot occur at high intraabdominal temperatures. The testes of most mammals undergo a marked caudal migration (descent) and come to lodge outside the body cavity in a *scrotum* where the temperature is several degrees lower than the abdominal temperature.

Terrestrial reproduction also necessitates a method of suppressing a free-swimming aquatic larva. Primitive reptiles remain egg laying, or oviparous, but suppress the larva by the evolution of a *cleidoic egg* (Fig. 9–3). This is an egg in which provision is made for all the requirements of the embryo, so

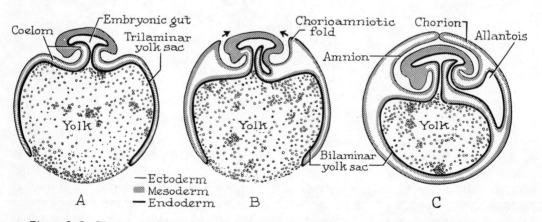

Figure 9–3 Diagrammatic sections of vertebrate embryos to show the extraembryonic membranes. *A*, Trilaminar yolk sac of a large yolked fish embryo: *B*, hypothetical derivation of the chorioamniotic fold of the amniote embryo from the superficial layers of the trilaminar yolk sac; *C*, extraembryonic membranes of an amniote embryo. (From Villee, Walker, and Barnes: General Zoology.)

that it can develop directly into a miniature adult capable of living on the land. The egg is supplied with a large store of yolk, which eventually becomes suspended in a *yolk sac.* As it descends the oviduct, *albumin,* or similar secretions which supply other metabolic needs, and a protective *shell* are added to the egg. The embryo itself early develops *extraembryonic membranes* that fulfill other needs. A protective *chorion* and a fluid-filled *amnion* (which provides a local aquatic environment) evolve from ectodermal and mesodermal layers that covered the yolk sac in such an animal as the dogfish. The yolk sac of amniotes is therefore *bilaminar* with a wall of just mesoderm and endoderm. Finally, a respiratory and excretory *allantois* evolves from the urinary bladder of amphibians.

Primitive prototherian mammals lay this type of egg, but in therian mammals this egg, minus its shell and albumin, is retained in the female reproductive tract, and a *placenta* evolves. In eutherian mammals, the placenta is simply a union of the chorion and allantois *(chorioallantoic membrane)* on the one hand with the uterine lining on the other. As will be seen, this mode of reproduction necessitates changes in the female reproductive tract, notably the evolution of a *uterus.*

STUDY OF THE EXCRETORY AND REPRODUCTIVE SYSTEMS

It is assumed that the major parts of the excretory and reproductive systems have been observed in previous dissections. In these exercises, the finer aspects will be examined and related to the more conspicuous parts. In studying these systems, you should not only dissect your own specimen, but also examine the dissection of a specimen of the opposite sex. Since someone else, in turn, will have to examine your specimen, make a particularly careful dissection. If possible, sexually mature specimens should be used.

(A) EXCRETORY SYSTEM

The *kidneys (renes)* of mammals, which are *metanephroi,* are located against the dorsal wall of the peritoneal cavity in a retroperitoneal position. Each is surrounded by a mass of fat *(adipose capsule),* which should be removed, and each is closely invested by a *fibrous capsule.* Notice that a kidney is bean shaped (Fig. 9-4 and 9-5). The indentation on the medial border is called the *hilus.* Carefully remove connective tissue from the hilus, and you will see the *renal artery* and *vein* entering the kidney, and, caudal to them, the *ureter* that drains the kidney.

Remove one of the kidneys and cut it longitudinally through the hilus in the frontal plane of the body. Study the half that includes the largest portion of the ureter. The hilus expands within the kidney into a chamber called the *renal sinus.* Pick away fat in the sinus in order to expose the blood vessels and the proximal end of the ureter. The sinus is largely filled with these structures, so its size is often not appreciated. If the vessels and ureter were removed, the space left would be the sinus. The portion of the ureter within the sinus is expanded and is termed the *renal pelvis* (Figs. 9–4 and 9–6). The substance of the kidney converges in the cat to form a single, nipple-shaped *renal papilla* which projects into the renal pelvis. In human beings there are many renal papillae, and the proximal portion of the renal pelvis is subdivided into chambers for them, called *calyces.* Notice that the substance of the kidney can be subdivided into a peripheral, light *cortex* and a deeper, darker *medulla.* The medullary substance of the cat constitutes one large *renal pyramid* whose apex is the renal papilla; in species with many papillae, there is a pyramid for each one.

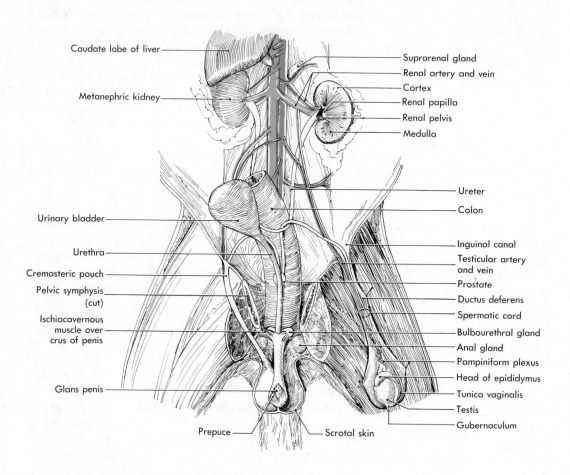

Caudate lobe of liver

Suprarenal gland
Renal artery and vein
Cortex
Renal papilla
Renal pelvis
Medulla

Metanephric kidney

Ureter
Colon

Urinary bladder

Inguinal canal
Testicular artery and vein
Prostate
Ductus deferens
Spermatic cord
Bulbourethral gland
Anal gland
Pampiniform plexus
Head of epididymus
Tunica vaginalis
Testis
Gubernaculum

Urethra
Cremasteric pouch
Pelvic symphysis (cut)
Ischiocavernous muscle over crus of penis

Glans penis

Prepuce
Scrotal skin

Figure 9–4 Ventral view of the urogenital system of a male cat. One kidney has been sectioned to show its internal structure. The pelvic canal has been cut open, and the cremasteric pouch has been dissected on the right side of the drawing.

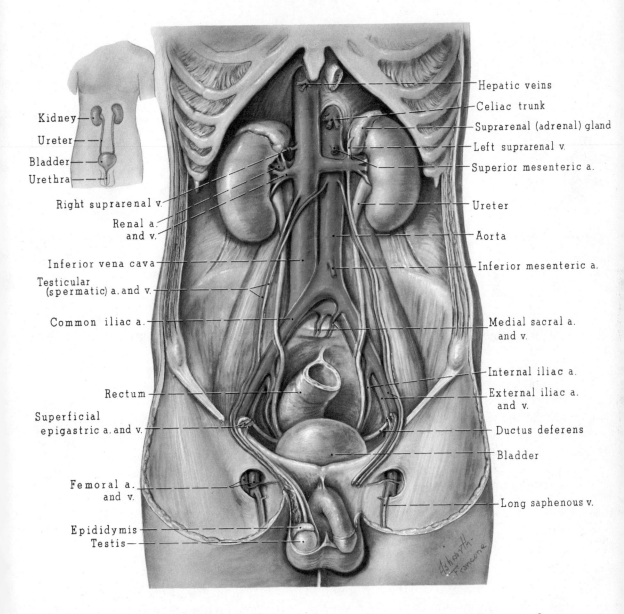

Kidney

Ureter

Bladder

Urethra

Right suprarenal v.

Renal a.
and v.

Inferior vena cava

Testicular
(spermatic) a. and v.

Common iliac a.

Rectum

Superficial
epigastric a. and v.

Femoral a.
and v.

Epididymis

Testis

Hepatic veins

Celiac trunk

Suprarenal (adrenal) gland

Left suprarenal v.

Superior mesenteric a.

Ureter

Aorta

Inferior mesenteric a.

Medial sacral a.
and v.

Internal iliac a.

External iliac a.
and v.

Ductus deferens

Bladder

Long saphenous v.

Figure 9–5 Anterior view of the urogenital system of a man. (From Jacob and Francone: Structure and Function in Man. W. B. Saunders Company.)

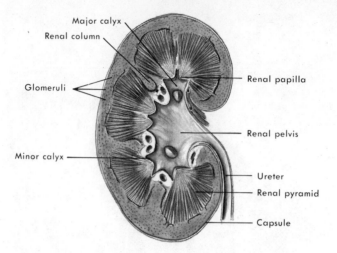

Major calyx
Renal column
Glomeruli
Minor calyx
Renal papilla
Renal pelvis
Ureter
Renal pyramid
Capsule

Figure 9–6 Frontal section of the human kidney. (From Dienhart: Basic Human Anatomy and Physiology. W. B. Saunders Company.)

Trace one of the ureters (Fig. 9–4). It extends caudally dorsal to the parietal peritoneum, and then turns ventrally in the lateral ligament of the bladder to enter the caudal part of the *urinary bladder.* As the ureter enters the lateral ligament of the bladder, it passes dorsal to the ductus deferens (male), or horn of the uterus (female). The urinary bladder itself is a pear-shaped organ with a broad, rounded, cranial end (its *vertex*) and a narrow, caudal end (its *body*). Cut open the bladder, and you may be able to see the points of entrance of the ureters in the dorsal wall of the bladder. Clean away connective tissue from around the bladder. It gradually narrows caudal to the entrance of the ureters and passes into the pelvic canal. This narrow passage, which carries urine to the outside, is called the *urethra.* The urethra begins just caudal to the entrances of the ureters. Its more caudal parts will be considered with the reproductive organs.

(B) MALE REPRODUCTIVE SYSTEM

For reasons stated in the introduction to the mammalian urogenital system, the testes of most mammals have undergone a descent so that they lie in a sac, the *scrotum,* outside the body cavity (Figs. 9–4 and 9–5). Carefully cut through the scrotal skin on each side, and separate the skin from the deeper layers of the scrotum. A dense layer of connective tissue containing some smooth muscle fibers, the *dartos tunic,* is closely associated with the skin and will come off with it. The dartos forms the septum between the left and right sides of the scrotum. Deeper layers of the scrotum take the form of a pair of cordlike sacs that extend caudad from the abdominal wall, cross the ventral surface of the pelvic girdle, and enter the skin sacs. The wall of these cordlike sacs is composed of the layers of the abdominal wall that were drawn down during the descent of the testes. A layer of *cremasteric muscle,* derived from the internal oblique musculature, is covered by the *external* and *internal spermatic fascia,* derived respectively from the aponeuroses of the external oblique and transversus muscles. It is difficult to differentiate these layers in a cat. Although considered to be a part of the scrotum, each sac may be called a *cremasteric pouch.* * The cremasteric pouches lie just beneath the skin, and should have been seen and saved when the pelvic canal was opened (p. 161). The testes lie within the caudal ends of the cremasteric pouches. This portion of each pouch is quite large in the cat and man, while the rest is a constricted tube, for the testes remain permanently in the scrotum. But all of the pouch is wide and of nearly uniform diameter in the rabbit, rat, and certain other mammals in which the testes move back and forth—into the scrotum during the breeding season, back into the abdomen the rest of the time. The reduction of the cremasteric muscle in the cat and its hypertrophy in the rabbit are correlated with whether the testes remain permanently in the scrotum or are migratory.

Leave the cremasteric pouch intact on one side, but cut open the other one along its ventral surface. Extend the cut from the caudal end of the pouch to the body wall, but do not cut through the body wall. Notice that the cremasteric pouch of the cat contains a cavity, the lumen of the *processus vaginalis,* or *tunica vaginalis* (Fig. 9–7). Pass a probe forward through this cavity near the body wall, and the probe will enter the peritoneal cavity. The processus vaginalis is a coelomic sac that descends with the testis. Thus the wall of the cremasteric pouch is lined with coelomic epithelium (the *parietal layer* of the processus vaginalis), and the structures within it (testis, epididymis, ductus deferens, testicular vessels and nerves) are covered with coelomic epithelium (the *visceral layer* of the processus vaginalis). The complex of ductus deferens and associated vessels and nerves, together with their covering of coelomic epithelium, is known as the *spermatic cord.* Notice that the cord is supported by a mesentery, a part of the *mesorchium,* passing from the dorsal wall of the pouch and that the most caudal structures in the pouch (testis and epididymis) and this part of the mesorchium are united to the caudal end of the pouch by a band of tissue called the *gubernaculum.* In many mammals, man included, only the distal portion of the processus vaginalis persists in the adult; the rest of the processus vaginalis atrophies. The human spermatic cord, therefore, is covered directly by the muscle and fascial layers of the cremasteric pouch.

The contents of the cremasteric pouch can now be examined in more detail. The *testis* is the relatively large, round body lying in the caudal part of the pouch (Figs. 9–4 and 9–5). Testicular blood vessels and nerves attach to its cranial end. Notice that the testicular artery coils upon itself before it reaches the testis and is closely invested by a venous network, the *pampiniform plexus,* formed by the testicular vein. It is possible that this is a mechanism that permits heat exchange from the arterial to the venous blood, which would help to keep testicular temperature low. The *epididymis* is a bandshaped structure closely applied to the surface of the testis. It can be divided into three regions—a *head* at the cranial end of the testis, a *body* on the lateral surface of the testis, and a *tail* at the caudal end of the testis. The head of the epididymis is functionally connected with the testis, and is made up of modified kidney tubules which have developed from the cranial mesonephric tubules (Fig. 9–2, *D*). The rest of

---

*The term scrotum properly includes all the layers separating the coelomic space in which each testis lies from the outside. It is useful, however, to have a separate term to designate the complex of tissues, apart from the skin and dartos tunic, that enfold each testis and its duct and vessels, because this complex is clearly defined and must be separated from the skin in order to see fully the anatomical relationships that result from the descent of the testes. I am following Prof. Gerard (in P.-P. Grassé: *Traité de Zoologie,* Masson et Cie., Paris, 1954, Vol. 12) in calling this part of the scrotum the cremasteric pouch.

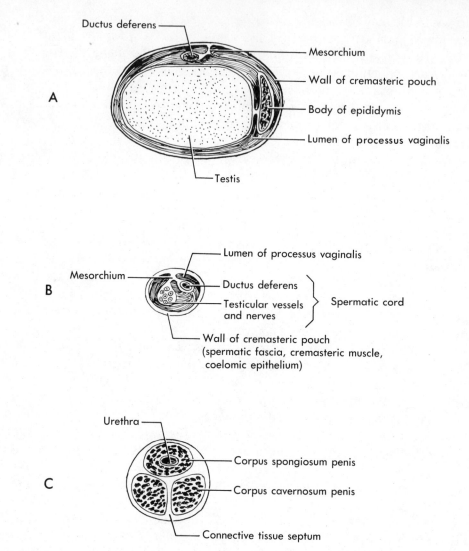

Figure 9–7 Semidiagrammatic, enlarged cross sections through parts of the reproductive system of a male cat. *A*, through the scrotum at the level of the testis, scrotal skin removed; *B*, through the middle of the cremasteric pouch; *C*, through the middle of the penis.

the epididymis consists of the highly convoluted ***ductus epididymidis*** (former cranial end of the archinephric duct) embedded in connective tissue.

The ***ductus deferens,*** the first part of which is somewhat convoluted, leaves the tail of the epididymis and, in company with the testicular vessels and nerves, ascends the cremasteric pouch and passes through the abdominal wall. The passage through the body wall is known as the ***inguinal canal.*** The canal is not long in the cat, its cranial end ***(internal inguinal ring)*** being the entrance into the peritoneal cavity; its caudal end ***(external inguinal ring),*** the attachment of the outermost layer of the wall of the cremasteric pouch (external spermatic fascia) to the aponeurosis of the external oblique muscle. In man the inguinal canal passes diagonally through the abdominal wall, hence is much longer.

Continue to follow the ductus deferens. It passes forward in the peritoneal cavity for a short distance, then loops over the ureter, and extends caudally into the pelvic canal between the urethra and large intestine. The ductus deferentes of opposite sides then converge and soon enter the urethra. The portion of the urethra distal to this union carries both sperm and urine. Various accessory sex glands, which secrete the seminal fluid, are associated with the ends of the ductus deferentes and adjacent parts of the urethra. In the cat, the two ductus deferentes enter the urethra independently, and a small *prostate* surrounds the point of entrance and the adjacent urethra. At the caudal end of the pelvic canal, a pair of *bulbourethral* (Cowper's) *glands* enter the urethral canal. These glands lie dorsal to a pair of processes, the *crura of the penis,* that extend from the base of the penis to the ischia. Each crus is covered by muscular tissue *ischiocavernosus muscle).* If the crura were not torn when the pelvic canal was opened, you may have to cut one now to see the bulbourethral glands clearly. In addition to these glands, man also has a *vesicular gland (seminal vesicle)* that joins the distal end of each ductus deferens.

The *penis* encloses the part of the urethra lying outside the pelvic canal. The free end of the penis, *glans penis,* lies in a pocket of skin called the *prepuce.* It is the prepuce that is removed by circumcision in many male infants. Cut open the prepuce to see better the glans and the opening of the urethral canal. A number of small spines are borne on the glans of the cat but these are absent in man. The rest of the penis is a

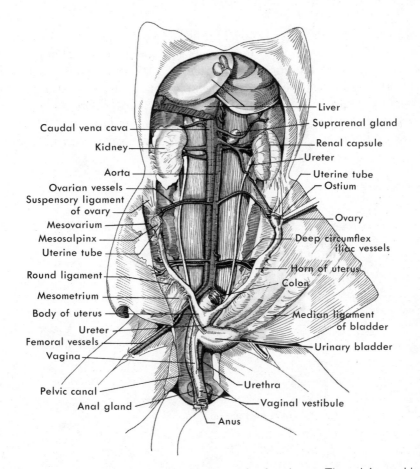

Figure 9–8 Ventral view of the urogenital system of a female cat. The pelvic canal has been cut open.

firm, cylindrical structure which should be exposed by removing the skin and surrounding loose connective tissue. Make a cross section of this portion of the penis and examine it with a hand lens. The urethral canal lies along the dorsal surface of the penis of the cat (if the organ is flaccid) embedded in a column of spongy tissue called the *corpus spongiosum penis* (Fig. 9-7, *C*). A pair of columns of spongy tissue separated by a septum, which is often indistinct, lie along the opposite surface and are surrounded by a ring of dense connective tissue. These columns are the *corpora cavernosa penis.* The glans penis is simply a caplike fold of the corpus spongiosum penis that covers the distal ends of the corpora cavernosa penis. The crura of the penis are the diverging proximal ends of the corpora cavernosa penis. The spongy tissue of which all these corpora consist is known as *erectile tissue,* and the spaces within it become filled with blood during erection. Make a cross section through the glans penis and look for a small bone, the *os penis* (baculum), that lies on one surface of the urethra and helps to stiffen this part of this penis. It is absent in man.

Before leaving the urogenital system, dissect beneath the skin on either side of the rectum near the anus and find a pair of round *anal glands.* They produce an odoriferous secretion that enters the anus and presumably is used as a pheromone for sexual attraction or stimulation.

(C) FEMALE REPRODUCTIVE SYSTEM

The *ovaries* are a pair of small, oval bodies (Figs. 9-8 and 9-10). In the adult they lie slightly caudal to the kidneys, for they have undergone a partial descent, and the metanephroi shift cranially during development. The small size of the mammalian ovaries is correlated with the uterine development of the embryos. Fewer eggs are produced, and the eggs do not contain much yolk. The eggs are microscopic, but you may see small vesicles, the *graafian follicles (folliculi vesiculosi),* each of which contains an egg, protruding on the surface of the ovary. The ovary protrudes into the body

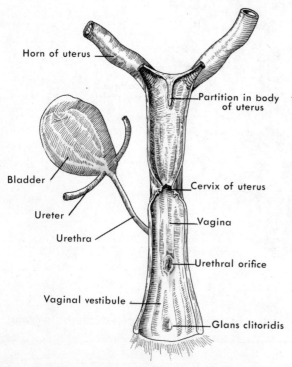

Horn of uterus

Partition in body of uterus

Bladder

Cervix of uterus

Ureter

Urethra

Vagina

Urethral orifice

Vaginal vestibule

Glans clitoridis

Figure 9-9 Dorsal dissection of the reproductive organs of a female cat to show the internal features of the uterus, vagina, and vaginal vestibule.

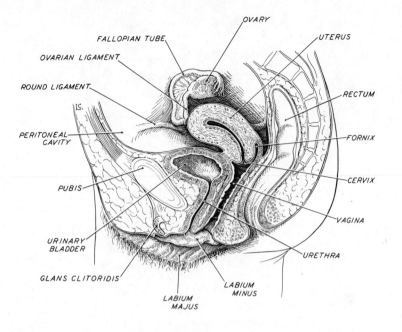

Figure 9–10 Sagittal section of the urogenital organs in the pelvic region of a woman. (From Turner and Bagnara: General Endocrinology. W. B. Saunders Company.)

cavity. When eggs are discharged from the follicles *(ovulation)* they pass through a part of the coelom to the entrance of the reproductive duct, which is not far away.

After ovulation the follicles are transformed into *corpora lutea,* which may also be seen protruding from the surface of the ovary, especially in pregnant specimens.

Typical oviducts are present in early mammalian embryos, but they differentiate into several regions during development, and their caudal ends fuse in varying degrees (Fig. 9–11). Thus, the adult reproductive tract is more or less Y-shaped. The cranial part of each wing of the Y forms a narrow, convoluted *uterine tube* (fallopian tube) lying lateral to the ovary (Fig. 9–8). Notice that a uterine tube curves over the front of the ovary and forms a hoodlike expansion *(infundibulum)* with fringed (fimbriated) lips. Spread open the lips, and you will see the coelomic opening of the tube, the *ostium,* which receives the eggs.

In the cat, the rest of each wing of the Y lies caudal to the ovary and forms a much wider tube—the *horn of the uterus.* It is very large in pregnant specimens, for the embryos develop within it. In a female human being the embryonic oviducts have fused to a greater degree. There are no uterine horns, and the uterine tubes enter a large, pear-shaped uterus of the simplex type (Figs. 9–10 and 9–11).

The ovary and reproductive tract of the female are supported by a mesentery known as the *broad ligament.* Often a great deal of fat lies within it. The portion of the broad ligament attaching to the uterus is the *mesometrium;* that attaching to the uterine tube, the *mesosalpinx;* and that attaching to the ovary, the *mesovarium* (Fig. 9–8). Pull the uterine horn toward the midline, thereby stretching the mesometrium. The mesenteric fold extending diagonally across the mesometrium from a point near the cranial end of the uterine horn to the body wall, and lying perpendicular to the broad ligament, is the *round ligament.* Notice that the round ligament attaches to the body wall at a point comparable to the location of the inguinal canal in the male. The

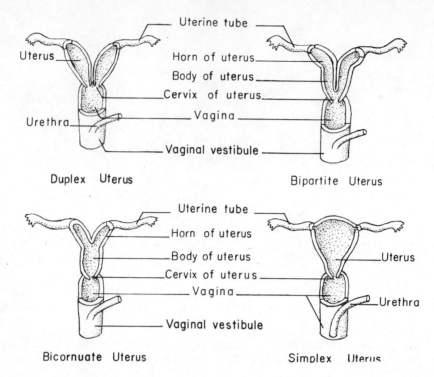

Figure 9-11 Diagrams to show the progressive fusion of the caudal ends of the oviducts in placental mammals. The uterus and part of the vagina have been cut open. The duplex type of uterus, in which the lower ends of the oviducts have united to form a vagina but the uteri remain distinct, is found in rodents and lagomorphs. In the bipartite uterus of carnivores, the lower ends of the uteri also have fused to form a median body from which uterine horns extend, but a partition is present in the body of the uterus. This partition is lost in the bicornuate uterus of ungulates. In the simplex uterus of primates, the uteri have completely united to form a large median body from which the uterine tubes arise. In primates, the vaginal vestibule also divides, so that the vagina and urethra open independently nearly on the body surface. (Modified after Wiedersheim: Comparative Anatomy of Vertebrates. The Macmillan Company.)

round ligament is the female counterpart of the male gubernaculum—a strand that plays an important role in the descent of the testis.

The two uterine horns of the cat converge cranial to the pelvic canal and enter a common median passage. This would be the stem of the Y, and it is formed in part by the fusion of the lower ends of the oviducts and in part by the division of the cloaca (see page 185). The cranial part of this median passage is the *body of the uterus.* The body is not long and soon leads into the *vagina,* which proceeds caudally through the pelvic canal, lying between the urethra and large intestine. Carefully separate these structures from each other and find the point where the vagina and urethra unite. The common passage from here on is the *vaginal vestibule.* It is a relatively long passage in quadrupeds. The comparable area in women is known as the *vulva,* but the vulva is very shallow, for vagina and urethra are independent nearly to the body surface. The opening of the vaginal vestibule, or vulva, is flanked by skin folds, the *labia,* but these are not conspicuous in quadrupeds.

Cut through the skin around the opening of the vaginal vestibule and completely free the vestibule and vagina from the rectum. A pair of round *anal glands* can be found by dissecting beneath the skin on the lateral surface of the rectum near the anus. These glands produce an odoriferous secretion that enters the anus and presumably is used as a pheromone for sexual attraction or stimulation. They are absent in human beings. Now open the median portion of the genital tract by making a longitudinal incision through its dorsal wall that extends from the vestibule to the horns of the uterus. Veer away from the middorsal line toward one of the uterine horns as you open the body of the uterus. A small bump may be seen in a pocket of tissue in the midventral line of the vaginal vestibule, near the orifice of the sinus. This is the *glans clitoridis,* which develops from the genital tubercle of the sexually indifferent stage of the embryo (Fig. 9–9).

More cranially in the vaginal vestibule, you will see the entrance of the urethra. The genital passage cranial to this union is the vagina, and it continues forward to the neck, or *cervix of the uterus.* In the cat, the cervix lies about halfway between the urethral orifice and the horns of the uterus and appears as a pair of folds constricting the lumen of the reproductive tract. The body of the uterus lies between the cervix and the horns. Notice that the cranial part of the body is subdivided into right and left sides by a vertical partition. The uterus of the cat is therefore bipartite (Fig. 9–11).

The origin of most of the mammalian male and female urogenital tracts from those of the lower vertebrates has been indicated. However, a consideration of the cloacal region and its fate in mammals has been deferred until the terminal portions of the urogenital passages could be studied in the mammal.

A *cloaca,* which receives material from both the digestive and urogenital tracts, is present in the lower vertebrates, in monotremes and in the embryos of the higher mammals. But it becomes divided and contributes to the intestinal and urogenital passages in the adults of the higher mammals. In an early, sexually indifferent, eutherian embryo, the cloaca consists of a chamber derived from the enlargement of the caudal end of the hindgut (Fig. 9–12, *A*). At first this endodermal cloaca is separated from the ectodermal invagination, the *proctodeum,* by a plate of tissue, but this plate soon breaks down and the proctodeum contributes to the cloaca.

The cloaca receives the intestine dorsally and the allantois ventrally. Even at an early stage (Fig. 9–12, *A*), the cranial portion of the cloaca is partly divided by a *urorectal fold* into a dorsal *coprodeum* receiving the intestine, and a ventral *urodeum* receiving the allantois, the ureters, and the archinephric ducts. Finally, a small *genital tubercle* is present on the ventral surface of the body cranial to the cloaca. This stage is similar to the cloaca of lower vertebrates except for the more ventral entrance of the urogenital ducts.

Later in the sexually indifferent period (Fig. 9–12, *B*), the urodeum and coprodeum become completely separated from each other by the urorectal fold and form the urogenital sinus and rectum, respectively. Oviducts now enter the front of the urogenital sinus beside the archinephric ducts, but the attachments of the ureters shift onto the allantois and developing urinary bladder.

In the subsequent differentiation of a male (Fig. 9–12, *D*), the constricted neck of the bladder (allantois) and the urogenital sinus form that portion of the urethra that is not included in the penis (segments 1 and 2). The archinephric ducts form the ductus deferentes and enter the urethra. Their point of entrance is a landmark that separates the portion of the urethra derived from the allantois from the portion derived from the urogenital sinus. The oviducts disappear in the male, although their point of entrance into the urethra may form a small sac (prostatic utricle) within the prostate. The genital tubercle enlarges to form the penis, and a groove on its ventral surface closes over, by the coming together of the *genital folds,* to form the penile portion of the urethra (segment 3).

In the subsequent differentiation of most female mammals (Fig. 9–12, *C*), the constricted neck of the bladder forms the entire urethra. The female urethra is thus comparable to only a small portion (the allantoic segment) of the male urethra. The urethra and the two oviducts, whose lower ends have fused to form the vagina and uterus, enter the urogenital sinus, which becomes the vaginal vestibule.

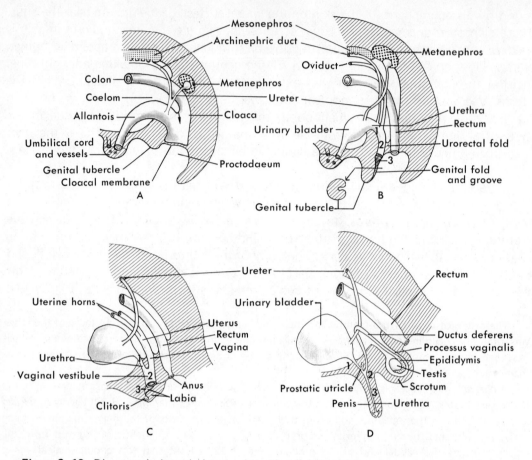

Figure 9–12 Diagrams in lateral view to show the division of the cloaca that occurs during the embryonic development of a placental mammal. *A*, an early sexually indifferent stage in which the division of the cloaca has just begun; *B*, a later sexually indifferent stage in which the cloaca has become divided into a dorsal rectum and ventral urogenital sinus; *C*, differentiation to the female condition; *D*, differentiation to the male condition. 1, 2, and 3 indicate comparable regions.

In most female mammals, the vestibule remains undivided. But in primates, it too becomes divided and continues the urethra and vagina nearly to the surface as separate passages. The archinephric ducts disappear, and the genital tubercle forms the small remnant known as the clitoris. The genital folds of the indifferent stage form the labia minora of a human being. The labia major are skin folds comparable to the scrotum of the male.

REPRODUCTION AND EMBRYOS

Eutherian mammals are viviparous; the embryos develop within the uterus and are born as miniature adults. If any of the specimens is pregnant, cut open the uterus and examine the embryos. Mammalian embryos produce the various extraembryonic membranes characteristic of amniotes (Fig. 9–3). Since the outermost membrane is the chorion, the whole complex of embryo and extraembryonic membranes is often called the *chorionic vesicle* (Fig. 9–13). Microscopic *chorionic villi* arise from the surface of the chorioallantoic membrane in eutherian mammals and penetrate or unite in various ways with the uterine lining. This combination of uterine lining and villi constitutes the

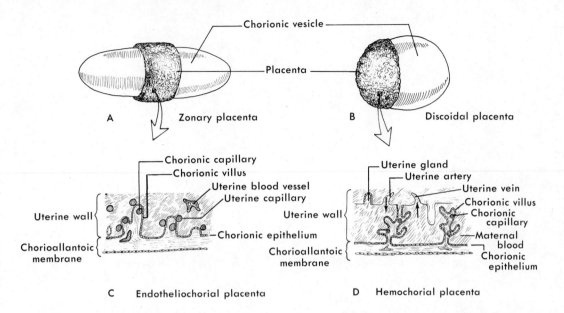

Figure 9–13 Diagrams of the cat and human placenta. *A*, chorionic vesicle of the cat; *B*, chorionic vesicle of a human being; *C*, enlarged detail of the cat placenta; *D*, enlarged detail of the human placenta.

placenta. In many mammals, including the cat and human being, the union of villi and uterine lining is intimate, and some of the uterine lining is discharged at birth. Such a placenta is said to be *deciduous,* in contrast to a *nondeciduous* placenta, in which the union is not intimate and maternal tissue is not discharged at birth.

Deciduous placentas have different shapes according to the distribution of the villi that make an intimate union with the uterine lining. In the cat, a belt-shaped band of villi unites with the uterine lining to form a definitive placenta of the *zonary* type. This band is easily seen on the surface of the chorionic vesicle. In human beings a disc-shaped patch of villi unites with the uterine lining, so the placenta is described as *discoidal.*

Still other terms describe the microscopic details of the union. In the cat, the surface epithelium lining the uterine portion of the placenta disappears. The chorionic villi penetrate the substance of the uterine lining and come in contact with the endothelial walls of the maternal capillaries to form an *endotheliochorial placenta.* The union is more intimate in the human *hemochorial placenta,* for maternal capillaries break down and chorionic villi are bathed in pools of maternal blood. To summarize, the cat has a chorioallantoic, deciduous, zonary, endotheliochorial placenta; we have a chorioallantoic, deciduous, discoidal, hemochorial placenta.

Cut open the chorionic vesicle and placenta, and you will see the embryo enclosed by the *amnion.* Open the amnion. The cord of tissue extending from the underside of the embryo to the chorionic vesicle and placenta is the *umbilical cord.* It contains the allantoic stalk, umbilical, or allantoic, blood vessels, and a vestige of the yolk sac stalk. Its surface is covered by the amnion.

Appendix I

WORD ROOTS

When studying anatomy for the first time, many students are confronted with numerous unfamiliar terms that must be mastered, because effective communication requires their use. Some understanding of the derivation of anatomical terms from their classical origins will help to fix their meaning and spelling in mind. The following list is a sampling of the more important word roots used in anatomy, and of the Greek (Gr.) and Latin (L.) from which they have been derived. Usually only the nominative singular is given for a noun, but the genitive (gen.), nominative plural (pl.), or diminutive (dim.) is included if sufficiently different or if it has formed the basis for an example cited. Only the infinitive for a verb is given, unless another form, such as the past participle (p.p.), is needed to recognize the root. Most of the examples cited are anatomical structures, but major groups of organisms mentioned in this book are included, and often a common word is cited to help fix the root in mind. This list of roots is far from inclusive; most familiar words for which the classical and the modern term and meaning are very similar have been omitted; e.g., auditory, cardiac, humerus, pharynx, ulna. Those whose appetites are whetted should consult standard dictionaries or such references as Jaeger (1955) for further study.

a — Gr. *a* or *an-*, prefix meaning without, not. (Agnatha, anapsid)
ab — L. *ab-*, prefix meaning away from. (Abduction)
abdom — L. *abdomen*, abdomen (probably from *abdo*, to conceal). (Abdomen, abdominal)
acanth — Gr. *akantha*, spine, thorn, (Acanthodii)
acetabul — L. *acetabulum*, vinegar cup (from *acetum*, vinegar). (Acetabulum)
acro — Gr. *akros*, topmost, extreme. (Acromion)
actin — Gr. *aktin*, ray. (Actinopterygii)
ad, af — L. *ad-* (may be changed to *af-* before certain words), prefix meaning motion toward. (Adduction, afferent)
af — See *ad*.
al — L. *ala*, wing. (Aliform, alisphenoid bone)
all — Gr. *allos*, different, other. (Allotheria)
allant — Gr. *allas*, gen., *allantos*, sausage. (Allantoic artery, allantois)
amphi — Gr. *amphi*, on both sides of, double. (Amphibian, amphicoelous, Amphioxus)
an — See *a-*.
anc — Gr. *ankon*, elbow, bend of the arm. (Anconeus muscle)
ante — L. *ante-*, prefix meaning in front of, previous to. (Antebrachium, anterior)
apo — Gr. *apo-*, prefix meaning away from. (Apophysis)
apsid — Gr. *apsis*, gen., *apsidos*, a loop. (Anapsid skull)
arachn — Gr. *arachne*, spider. (Arachnid, arachnoid membrane)
archi — Gr. *archi-*, prefix meaning primitive. (Archinephros, archipallium)
aryten — Gr. *arytaina*, ladle. (Arytenoid cartilage)
atri — L. *atrium*, entrance room. (Atrium)
aur — L. *auris*, dim., *auricula*, ear. (Auricle)

av — L. *avis*, pl., *aves*, bird. (Aves, aviary)

bi, bin — L. *bi-, bin-*, prefix meaning two. (Biceps muscle, binocular, biped)

bi, bio — L. *bios*, life. (Amphibian, biology)

brachi — Gr. *brachion*, upper arm. (Antebrachium, brachialis muscle, brachium pontis)

branchi — Gr. *branchion*, gill. (Branchial, Elasmobranchii)

bull — Gr. *bulla*, a bubble, seal. (Tympanic bulla)

caec — See *cec-*.

calc — L. *calx*, gen., *calcis*, lime, heel. (Calcaneus, calcify, calcium)

caly — Gr. *kalyx*, cup. (Renal calyx)

capit — L. *caput*, gen., *capitis*, dim., *capitulum*, head. (Capitulum, caput)

cardinal — L. *cardinalis*, principal, red. (Cardinal vein)

carn — L. *caro*, gen., *carnis*, flesh. (Carnal, carnivore, trabeculae carneae)

carotid — Gk. *karotides*, large neck arteries (from *karoun*, to stupefy). (Carotid artery)

carp — Gr. *karpos*, wrist. (Carpal, carpus)

caud — L. *cauda*, tail. (Caudal, caudofemoralis muscle)

cav — L. *cavus*, hollow. (Cavity, vena cava)

cec — L. *caecum*, blind intestine. (Cecum, ileocecal valve)

cel — See *coel*.

ceph — Gr. *kephale*, head. (Cephalic, holocephalic, triceps)

cer — Gr. *keras*, horn. (Ceratobranchial, keratin)

cervic — L. *cervix*, gen., *cervicis*, neck. (Cervical, cleidocervicalis muscle, uterine cervix)

chiasm — Gr. *chiasma*, cross. (Optic chiasma)

choan — Gr. *choane*, funnel. (Choana)

chol — Gr. *chole*, bile. (Ductus choledochus)

chondr — Gr. *chondros*, cartilage. (Chondrichthyes, chondrify, chondrocranium)

chord — L. *chorda*, string. (Chordae tendineae, Chordata, notochord)

chorio — Gr. *khorion*, skin, afterbirth. (Chorion, choriod layer of eyeball, choroid plexus)

choro — See *chorio*.

ciner — L. *cinereus*, the color of ashes (from *cinis*, ashes). (Incinerate, tuber cinereum)

cingul — L. *cingulum*, girdle. (Cingulate cortex)

clav —L. *clavis*, dim., *clavicula*, key; also, the clavicle. (Clavicle, subclavian artery)

cleid — Gr. *kleis*, gen., *kleidos*, key; also, the clavicle. (Cleidoic egg, cleidomastoid muscle)

cnem — Gr. *kneme*, shin or tibia. (Gastrocnemius muscle)

coel — Gr. *koilos*, hollow. (Acoelous, coelom)

coll — L. *collis*, dim., *colliculus*, hill. (Superior colliculus)

commissur — L. *commissura*, point of union of two things. (Rostral commissure of brain)

condyl — Gr. *kondylos*, articulatory prominence at a joint, a knuckle. (Epicondyle, occipital condyle)

conju — L. *conjungere*, p.p., *conjunctus*, to join together. (Conjugal, brachium conjunctivum, conjunctiva of eye)

corac — Gr. *korax*, gen., *korakos*, a raven or crow. (Coracobrachialis muscle, coracoid [resembles a crow's beak])

corn — L. *cornu*, horn. (Cornea, cornify)

coron — L. *corona*, crown, or something curved. (Coronary arteries, coronoid process)

cort — L. *cortex*, gen., *corticis*, bark. (Cerebral cortex, corticospinal tract)

cost — L. *costa*, rib. (Costal cartilage, intercostal space)

cox — L. *coxa*, hip bone. (Coxal)

cran — Gr. *kranion*, skull. (Cranial)

cre — Gr. *kreas*, flesh. (Cremasteric muscle, pancreas)

cribr — L. *cribrum*, sieve. (Cribriform plate)

cric — Gr. *krikos*, ring. (Cricoid cartilage)

crus — L. *crus*, pl., *crura*, lower leg or shank. (Abductor cruris muscle, crural fascia, crus penis) *cun* — L. *cuneus*, wedge. (Cuneiform bone, funiculus cuneatus)

cut — L. *cutis*, skin. (Cutaneous, cuticle, cutis)

cyclo — Gr. *kyklos*, circle. (Cyclic, Cyclostomata)

cyst — Gr. *kystis*, bladder. (Cyst, cystic duct)

cyt — Gr. *kytos*, hollow vessel, cell. (Lymphocyte, osteocyte)

daeum — See *deum*.

delt — Gr. *delta*, fourth letter of the Greek alphabet, Δ. (Deltoid muscle)

dendr — Gr. *dendron*, tree. (Dendrite of a nerve cell)

dens, dent — L. *dens*, gen., *dentis*, tooth. (Dens, dentary bone, dentine, dentist)

derm — Gr. *derma*, skin, leather. (Dermal bone, dermis, ostracoderm)

deum — Gr. *hodaios*, on the way. (Proctodeum, stomodeum)

di — Gr. *di-*, prefix meaning two. (Diapsid reptile, digastric muscle)

dia — Gr. *dia-*, prefix meaning across, between. (Di[a]encephalon, diapophysis, diaphragm)

didym — Gr. *didymos*, testicle. (Epididymis)

digit — L. *digitus*, finger. (Digit, digitigrade, extensor digitorum muscle)

don — See *odon*.

dors — L. *dorsum*, back. (Dorsal, longissimus dorsi)

duct — L. *ductus*, a leading (from *ducere*, to lead). (Abduction, duct, oviduct)

duoden — New L. *duodenum*, the first part of the intestine, which is about twelve finger-breadths long (a contraction of *intestinum duodenum digitorum*, from *duodeni*, twelve each). (Duodenal, duodenum)

dur — L. *durus*, hard. (Durable, dura mater)

ef — See *ex*.

elasm — Gr. *elasma*, a thin plate. (Elasmobranchii)

encephal — Gr. *enkephalos*, brain (from *en* + *kephale*). (Diencephalon, encephalitis, mesencephalon)

endo, ento — Gr. *endon*, *entos*, within. (Endoderm, endostyle, endotherm, entoderm, entotympanic bone)

enter — Gr. *enteron*, intestine. (Enteron, mesentery)

epi — Gr. *epi-*, prefix meaning upon. (Epicondyle, epidermis, epididymis, epithalamus)

ethm — Gr. *ethmos*, sieve. (Ethmoid bone)

eu — Gr. *eu-*, prefix meaning good, true. (Eutheria)

ex — L. *ex-*, *ef-*, prefix meaning out or away from. (Efferent neuron, extrinsic muscle)

falc — L. *falx*, gen., *falcis*, sickle. (Falciform ligament, falx cerebri)

fasc — L. *fascia*, band, bandage. (Fascia)

fasci — L. *fascis*, dim., *fasciculus*, bundle. (Fasciculus cuneatus)

fenestr — L. *fenestra*, window. (Fenestra ovalis, temporal fenestra)

fer — L. *ferre*, to carry. (Afferent neuron, deferent duct, efferent artery)

fibul — L. *fibula*, buckle, pin. (Fibula)

fid — See *fiss*.

fil — L. *filum*, dim., *filamentum*, thread. (Filamentous, filiform papillae)

fimb — L. *fimbria*, fringe. (Fimbria of hippocampus)

fiss — L. *findere*, p.p., *fissus*, to split. (Fissure, multifidus muscle)

flex — L. *flectere*, p.p., *flexus*, to bend. (Circumflex iliac artery, flexion)

foli — L. *folium*, leaf. (Cerebellar folia, foliage, foliate papillae)

foram — L. *foramen*, an opening. (Foramen magnum)

form — L. *forma*, shape, rule. (Cribriform plate, falciform ligament, formula, lentiform nucleus)

forn — L. *fornix*, a vault or arch; also, a brothel. (Fornicate, hippocampal fornix)

foss — L. *fossa*, a ditch (from *fodere*, p.p., *fossus*, to dig). (Fossil, fossorial, mandibular fossa)

fron — L. *frons*, gen., *frontis*, forehead. (Frontal bone)

fun — L. *funis*, dim., *funiculus*, cord, rope. (Funiculi of spinal cord, funicular) *fund* — L. *fundus*, bottom. (Gastric fundus)

fung — L. *fungus*, mushroom. (Fungiform papilla, fungus)

gang — Gr. *ganglion*, a swelling, tumor. (Ganglion)

gastr — Gr. *gaster*, stomach. (Digastric muscle, gastric glands, gastrocnemius muscle)

gen — L. *genus*, race or kind (from *genere*, to beget). (Genesis, genital, genus)

gen — L. *genu*, knee. (Geniculate bodies, genioglossus muscle)

glen — Gr. *glene*, cavity or socket. (Glenoid cavity)

gloss — Gr. *glossa*, tongue. (Genioglossus muscle, hypoglossal nerve)

glut — Gr. *gloutos*, buttock. (Gluteus muscle)

gnath — Gr. *gnathos*, jaw. (Agnatha, gnathostome)

gracil — L. *gracilis*, slender. (Fasciculus gracilis, gracile, gracilis muscle)

gubernacul — L. *gubernaculum*, a rudder (from *gubernare*, to steer). (Gubernaculum of testis, gubernatorial)

gyr — Gr. *gyros*, circle, round. (Cerebral gyri, gyrate)

haben — L. *haben*, dim., *habenula*, a strap. (Habenula of epithalamus)

ham — L. *hamus*, dim., *hamulus*, a hook. (Hamate bone, hamulus of pterygoid bone)

hepa — Gr. *hepar*, gen., *hepatos*, liver. (Hepatic duct)

hippo — Gr. *hippos*, horse. (Hippocampus)

hol — Gr. *holos*, whole, entire. (Holocephalic, holonephros, Holostei)

homo — Gr. *homos*, alike, the same. (Homology).

hyo — Gr. *hyoeides*, in the form of the letter upsilon (U-shaped). (Hyoglossus muscle, hyoid bone)

hyper — Gr. *hyper-*, prefix meaning above, over. (Hyperactive, hypertrophy)

hypo — Gr. *hypo-*, prefix meaning under, beneath. (Hypoglossal nerve, hypophysis, hypothalamus)

ichthy — Gr. *ichthys*, gen., *ichthyos*, fish. (Osteichthyes)

ile, ili — L. *ileum* or *ilium*, the groin. (Ileum, iliac, ilium)

incu — L. *incus*, an anvil. (Incus of ear)

ineum — Gr. *inan*, to excrete. (Perineum)

infra — L. *infra*, below. (Infraorbital foramen, infraspinatus muscle)

infundibul — L. *infundibulum*, a funnel. (Infundibulum of the uterine tube)

inguin — L. *inguen*, groin, *inguinalis*, pertaining to the groin. (Inguinal canal)

inter — L. *inter*, between. (Intercostal muscle, internuncial neuron, interparietal bone)

irid — Gr. *iris*, gen., *iridos*, rainbow. (Iridescent, iris of the eye)

ischi — Gr. *ischion*, hip. (Ischial, ischium)

jejun — L. *jejunus*, fasting, empty.(Jejunum [so called because it is usually found to be empty in dissection])

jugu — L. *jugum*, dim., *jugulum*, collarbone. (Jugular vein)

labyrinth — Gr. mythology *Labyrinthos*, the maze in which the Minotaur was confined. (Labyrinthodontia)

lachrim — L. *lacrima*, tear. (Lacrimal duct)

lamin — L. *lamina*, dim., *lamella*, layer or thin plate. (Lamellar bone, lamina terminalis, laminate)

lat — L. *latus*, gen., *lateris*, side. (Fascia lata, lateral, latissimus dorsi muscle)

len — L. *lens*, gen., *lentis*, a lentil. (Lens, lentiform nucleus)

lien — L. *lien*, spleen. (Gastrolienic ligament, lienogastric artery)

limb — L. *limbus*, edge, border. (Limbic lobe)

lingu — L. *lingua*, dim., *lingula*, tongue. (Lingual nerve, sublingual gland)

liss — Gr. *lissos*, smooth. (Lissamphibia)

lumb — L. *lumbus*, loin. (Lumbar artery)

lun — L. *luna*, moon. (Lunate bone, lunatic, semilunar valve)

magn — L. *magnus*, great. (Foramen magnum)

mam, mamma — L. *mamma*, gen., *mammalis*, dim., *mamilla*, breast. (Mamillary body, Mammalia) *man* — L. *manus*, hand. (Manual, manubrium, manus)

marsup — L. *marsupium*, purse, pouch. (Marsupialia, marsupium)

mass — Gr. *maseter*, a chewer. (Masseter muscle)

mast — Gr. *mastos*, breast. (Mastoid process)

meat — L. *meatus*, passage. (Acoustic meatus)

mediastin — L. *mediastinus*, median (from *medius*, middle). (Mediastinum)

medull — L. *medulla*, marrow. (Medulla oblongata, renal medulla)

mes — Gr. *mesos*, middle. (Mesencephalon, mesentery, mesoderm, mesonephros)

met, meta — Gr. *meta*, beside, after. (Metacarpal, metacromion, metanephros, Metatheria)

metr — Gr. *metra*, uterus. (Endometrium, mesometrium)

mon — Gr. *monas*, single. (Monarch, Monotremata)

mult — L. *multus*, many. (Multifidus muscle, multiply)

myel — Gr. *myelos*, marrow, spinal cord. (Myelencephalon, myelin)

myl — Gr. *myle*, millstone, molar. (Mylohyoid muscle)

myo — Gr. *mys*, muscle. (Myology, myomere, myotome)

nar — L. *naris*, pl., *nares*, external nostril. (Naris)

nas — L. *nasus*, nose. (Nasal bone)

nav — L. *navis*, dim., *navicula*, ship. (Navicular bone, navy)

neo — Gr. *neos*, new, recent. (Neo-Darwinism, neopallium)

neph — Gr. *nephros*, kidney. (Mesonephros, nephron)

neuro — Gr. *neuron*, tendon, nerve. (Neurocoele, neuron)

nict — L. *nictare*, to wink. (Nictitating membrane)

nunc — L. *nuncius*, messenger. (Internuncial neuron, nuncio)

obturat — L. *obturare*, p.p., *obturatus*, to close by stopping up. (Obturator foramen)

occipit — L. *occiput*, gen., *occipitis*, the back of the head. (Occipital bone)

ocul — L. *oculus*, eye. (Bulbus oculi, orbicularis oculi muscle)

oid — New L. *-oid* (from Gr. *o* + *eidos*, form), suffix that indicates resemblance to. (Arachnoid, diploid, pterygoid, sphenoid)

olecran — Gr. *olekranon*, elbow tip. (Olecranon)

olf — L. *olfacere*, to smell. (Olfactory organ)

om — Gr. *omos*, shoulder. (Metacromion, omohyoid muscle)

oment — L. *omentum*, membrane. (Greater omentum)

oo — Gr. *oon*, egg. (Epoophoron, oogenesis)

opt — Gr. *optikos*, pertaining to sight. (Optic nerve)

orb — L. *orbis*, dim., *orbiculus*, circle. (Orbicularis oculi muscle, orbit)

orch — Gr. *orchis*, testicle. (Mesorchium, orchid [from shape of root])

os, oss — L. *os*, gen., *ossis*, dim., *ossiculum*, bone. (Ossicle, ossify)

ost — Gr. *osteon*, bone. (Osteichthyes, teleost)

ostrac — Gr. *ostrakon*, shell. (Ostracoderm)

ot — Gr. *otikos*, pertaining to the ear. (Otic capsule, parotid gland)

ov — L. *ovum*, egg. (Ovary, oviduct, oviparous)

palae, pale — Gr. *palaios*, ancient. (Paleontology, paleopallium)

palat — L. *palatum*, roof of the mouth. (Palate, palatine bone)

palli — L. *pallium*, a Roman cloak. (Neopallium)

palpeb — L. *palpebra*, eyelid. (Levator palpebrae superioris muscle)

pampin — L. *pampinus*, tendril. (Pampiniform plexus)

pan — Gr. *pan*, all. (PanAmerican, pancreas)

para — Gr. *para*, beside. (Parapophysis, paradidymis, parasympathetic nerve)

pariet — L. *paries*, gen., *parietis*, wall. (Parietal peritoneum)

pat — L. *patina*, dim., *patella*, dish or plate. (Patella bone)

path — Gr. *pathetikos*, sensitive, liable to suffer (from *pathos*, suffering). (Pathology, sympathetic nervous system) *pect* — L. *pectus*, chest. (Pectoral girdle, pectoralis muscle)

pectin — L. *pecten*, comb. (Pectinate muscles of heart, pectineus muscle)

ped — L. *ped*, gen., *pedis*, dim., *pedunculus*, foot. (Bipedal, cerebral peduncle, pes)

pelluc — L. *pellucidis*, to shine through. (Pellucid, septum pellucidum)

pelv — L. *pelvis*, basin. (Pelvic girdle, renal pelvis)

peri — Gr. *peri-*, prefix meaning around. (Perineum, periosteum, peritoneum)

peron — Gr. *perone*, pin, fibula. (Peroneal nerve, peroneus muscle)

petr — Gr. *petros*, rock. (Petrify, petrosal bone)

phalang — Gr. *phalanx*, gen., *phalangos*, a line of soldiers. (Phalanges)

phor — Gr. *phoros,* from *pherein*, to bear. (Epoophoron, paroophoron)

phragm — Gr. *phragma*, fence, partition. (Diaphragm)

phren — Gr., *phren*, diaphragm, mind. (Phrenic nerve, phrenology)

phys — Gr. *physis*, a growth. (Apophysis, hypophysis, symphysis)

pia -- L. *pia*, tender. (Pia mater)

pin — L. *pinus*, pine tree. (Pineal gland)

pir — L. *pirum*, pear. (Piriform lobe, piriformis muscle)

pis — L. *pisum*, pea. (Pisiform bone)

plac — Gr. *plax*, gen., *plakos*, flat plate. (Placoderm, placoid scale)

plant — L. *planta*, sole of the foot. (Plantaris muscle, plantigrade)

platy — Gr. *platys*, flat or broad. (Duckbill platypus, platysma muscle)

pleur — Gr. *pleura*, side, rib. (Pleural cavity, pleurapophysis)

poll — L. *pollex*, gen., *pollicis*, thumb. (Adductor pollicis muscle)

pon — L. *pons*, gen., *pontis*, bridge. (Brachium pontis, pons)

poplit — L. *poples*, gen., *poplitis*, knee joint. (Popliteal fossa, popliteus muscle)

post — L. *post*, after, behind. (Posterior, postorbital process)

pre — L. *prae-*, prefix meaning before, in front. (Precaval vein, premaxillary bone)

prepu — L. *praeputium*, the foreskin. (Prepuce)

prim — L. *primus*, first. (Primates, primitive)

pro — Gr. *pro-*, prefix meaning before, in front. (Pronephros, prosencephalon)

proct — Gr. *proktos*, the anus. (Proctodeum)

prostat — Gr. *prostates*, one who stands before. (Prostate [stands before the bladder])

prot — Gr. *protos*, first. (Protoplasm, Prototheria)

psoa — Gr. *psoa*, a muscle of the loin. (Psoas major muscle)

pteryg — Gr. *pterygion*, gen., *pterygos*, wing, fin. (Actinopterygii, pterygoid process)

pub — L. *puber*, young adult. (Puberty, pubic hair, pubis)

pudend — L. *pudendum*, external genitals (from *pudere*, to be ashamed). (Pudendal artery, pudendum)

pulmo — L. *pulmo*, gen., *pulmonis*, lung. (Pulmonary artery)

pulvi — L. *pulvinus*, cushion. (Pulvinar)

pylor — Gr. *pylorus*, gate keeper. (Pyloric region of stomach, pylorus)

quad — L. *quadrare*, p.p., *quadratus*, to make square (from *quadrus*, square). (Quadrate bone, quadriceps femoris muscle)

radi — L. *radius*, ray, spoke. (Corona radiata, radius)

rect — L. *rectus*, straight. (Rectus abdominis muscle, rectus femoris muscle)

ren — L. *renes*, kidneys. (Renal artery)

rept — L. *reptare*, p.p., *reptum*, to creep. (Reptilia)

retinacul — L. *retinaculum*, band, holdfast. (Extensor retinaculum)

rhin — Gr. *rhis*, gen., *rhinos*, nose. (Rhinal fissure, rhinarium, rhinoceros)

rhomb — Gr. *rhombus*, parallelogram with oblique angles and unequal adjacent sides. (Rhombencephalon, rhomboideus muscle)

rug — L. *ruga*, fold. (Gastric rugae)

sacr — L. *os sacrum*, sacred bone (from *sace*, sacred). (Sacral nerve, sacrum) *sagitt* — L. *sagitta*, arrow. (Sagittal plane)

salp — Gr. *salpinx*, trumpet. (Mesosalpinx)

sarco — Gr. *sarx*, gen., *sarkos*, flesh. (Sarcophagus, Sarcopterygii)

sartor — Gr. *sartor*, tailor. (Sartorius muscle)

scal — L. *scala*, ladder. (Scala vestibuli)

scale — Gr. *skalenos*, a triangle with three unequal sides. (Scalenus muscle)

scaph — Gr. *skaphe*, bowl, boat. (Scaphoid bone)

scler — Gr. *skleros*, hard. (Sclera of eyeball, sclerous)

scrot — L. *scrotum*, pouch, scrotum. (Scrotum)

semi — L. *semi-*, prefix meaning partly, half. (Semicircular canal, semispinalis muscle, semitendin-osus muscle)

serr — L. *serra*, saw. (Serrated, serratus ventralis muscle)

sesa — Gr. *sesame*, seed of the sesame plant. (Sesamoid bone)

sin — L. *sinus*, a curve, hollow, bay. (Sinus venosus)

sole — L. *solea*, sandal. (Sole of the foot, soleus muscle)

som — Gr. *soma*, gen., *somatos*, body. (Somatic, somatopleure)

sperm — Gr. *sperma*, gen., *spermatos*, seed, semen. (Sperm, spermatic fascia)

sphen — Gr. *sphen*, wedge. (Sphenoid bone)

spin — L. *spina*, thorn or spine. (Erector spinae muscle, spinalis muscle, spine)

splanchn — Gr. *splanchnos*, viscus. (Splanchnic nerve, splanchnopleure)

splen — Gr. *splenion*, bandage.(Splenius muscle)

squam — L. *squama*, scale. (Squamata, squamosal bone, squamous epithelium)

stap — L. *stapes*, stirrup. (Stapes bone)

stern — Gr. *sternon*, breast, chest. (Sternebrae, sternomastoid muscle, sternum)

stom — Gr. *stoma*, gen., *stomatos*, mouth. (Cyclostomata, stomodeum)

strat — L. *stratum*, layer. (Stratum corneum)

stri — L. *striare*, p.p., *striatus*, to make furrows or stripes. (Corpus striatum, striated)

styl — Gr. *stylos*, pillar, stalk. (Styloid process, stylomastoid foramen)

sub — L. *sub-*, prefix meaning beneath, below. (Subscapular fossa)

sulc — L. *sulcus*, furrow. (Sulcus of brain)

supra — L. *supra-*, prefix meaning above, beyond. (Suprarenal gland, supraspinatus muscle)

sym — See *syn.*

syn — Gr. *syn-*, *sym-*, prefix meaning together or with. (Symbiosis, sympathy, symphysis, synapsid skull)

tal — L. *talus*, ankle. (Talus bone)

tars — Gr. *tarsos*, flat surface, sole of foot. (Metatarsal bone, tarsus)

tect — L. *tectum*, roof. (Optic tectum)

tel — Gr. *tele*, far off, distant. (Telencephalon, teleost, television)

tempor — L. *tempus*, gen., *temporis*, time. (Temple of the head, temporal bone)

ten — L. *tendere*, to stretch or extend. (Extend, neoteny, tendon, tensor muscle)

ter — L. *teres*, round, smooth. (Teres major muscle, ligamentum teres)

test — L. *testis*, a witness (originally an adult male). (Testify, testicular, testis)

thalam — Gr. *thalamos*, inner chamber. (Epithalamus, thalamus)

ther — Gr. *therion*, wild beast. (Eutheria, Prototheria)

thorac — Gr. *thorax*, gen., *thorakos*, chest. (Thoracic, thorax)

thym — Gr. *thymos*, the thymus (akin to *thymon*, thyme or a lump resembling a bunch of thyme). (Thymine, thymus)

thyr — Gr. *thyreos*, an oblong shield. (Thyrohyal muscle, thyroid cartilage)

ton — Gr. *tonos*, something stretched. (Peritoneum, tone, tonus)

tract — L. *trahere*, p.p., *tractus*, to pull or draw. (Protractor muscle, retractor)

trans — L. *trans*, across, beyond. (Transcend, transverse section)

trem — Gr. *trema*, hole. (Monotremata, pretrematic nerve branch) *tri* — L. *tri*, three (Triceps muscle, tricuspid valve)

triquetr — L. *triquetrus*, triangular. (Triquetrum bone)

trochlea — L. *trochlea*, pulley. (Trochlea on humerus, trochlear nerve)

tuber — L. *tuber*, dim., *tuberculum*, bump, lump. (Tuber cinereum, tuberculum of a rib)

turb — L. *turbo*, gen., *turbinis*, a spinning thing. (Turbinate bone, turbine)

tympan — L. *tympanum*, drum. (Tympanic membrane)

ur — L. *urina*, urine. (Urea, urinary, urogenital)

vag — L. *vagus*, wandering, undecided. (Vague, vagus nerve)

vagin — L. *vagina*, sheath. (Vagina, vaginal ligaments, processus vaginalis)

vast — L. *vastus*, large area, immense. (Vast, vastus lateralis muscle)

ventr — L. *venter*, dim., *ventriculus*, belly, womb. (Ventral, ventricle)

vesic — L. *vesica*, dim., *vesicula*, bladder. (Vesica fellea, vesicular gland)

visc — L. *viscus*, pl., *viscera*, an entrail. (Visceral)

vitell — L. *vitellus*, yolk. (Vitelline veins)

vitr — L. *vitrum*, glass. (Vitreous body of eye, vitrify)

vom — L. *vomer*, plowshare (cutting blade of a plow). (Vomer bone.)

vor — L. *vorare*, to devour. (Carnivore, voracious)

xiph — Gr. *xiphos*, sword. (Xiphisternum)

zyg — Gr. *zygon*, yoke, union, pair. (Azygos vein, zygapophysis, zygomatic arch, zygote)

Appendix II

THE PREPARATION OF SPECIMENS

It is common practice to purchase specimens from biological supply houses, but students are frequently interested in the way they are prepared, and it is sometimes necessary to prepare a specimen. The following directions are presented with these needs in mind. Only the simpler and more common procedures are described. Further details can be found in such a book as Wilder and Gage, *Anatomical Technology,* or can be obtained by writing to a biological supply house.* Directions for preparing many interesting and valuable teaching demonstrations can be found in Hildebrand's *Anatomical Preparations.*

1. KILLING THE SPECIMEN

After a specimen is caught, it must, of course, be killed. And, incidentally, cats should be obtained from the proper sources. Long years have passed since Wilder and Gage could write, "There is usually no difficulty in taking a cat when it is wanted. Such as will not come when called may be secured by means of a strong net. . . ."! Mammals may be killed by chloroforming or etherizing in a tight container, but injection of a barbiturate, such as sodium pentobarbital, is more humane if barbiturates can be obtained from the institutional physician or a local veterinarian. One may also gas a mammal by putting the specimen in a container and inserting a hose from a gas line. Turn the gas off when the specimen becomes quiet, but leave the specimen in the container for 5 to 10 minutes.

2. PRESERVING SPECIMENS

Many animals can be preserved simply by cutting open the body cavity and immersing the specimen in a solution of alcohol or formalin. The preserving solution can also be injected into the body cavity and into the larger muscle masses. For most vertebrates, use a 70 per cent solution of alcohol, or an 8 to 10 per cent solution of formalin. Commercial formalin is a saturated, aqueous solution (40 per cent) of formaldehyde gas. An 8 per cent formalin solution is a mixture of 8 parts of commercial formalin and 92 parts of water. Better results can be obtained on large specimens, however, by injecting the preserving fluid into the circulatory system—a process called embalming. After a cat has been killed, it should be tied out on a spreading board in the position in which one desires it to harden. Then make an incision through the skin on the medial surface of the thigh to expose the femoral artery. The artery is smaller and lighter in color than the accompanying vein.

---

*I am indebted for much of the following information to various "Turtox Service Leaflets," published by the General Biological Supply House, Chicago, Illinois.

Mobilize the artery and slip a thread beneath it. Then make a V-shaped cut halfway through the vessel with a pair of scissors. Insert a 16- or 18-gauge injection needle into the vessel, and tie it in. Force a wad of cotton into the animal's mouth to prevent the embalming solution from leaking.

An embalming solution that is often used is made according to the following formula:

Phenol (Carbolic acid, melted crystals)	5 parts
Formalin (40%)	5 parts
Glycerin	5 parts
Water	85 parts

The formalin is the chief preservative; the carbolic acid is a disinfectant and also helps to preserve the color of the tissues, and gives a sanitary odor to the specimen; the glycerin helps to maintain the pliability of the tissues.

The solution can be injected with a gravity bottle or with a large hand syringe. Inject sufficient fluid so that the specimen appears bloated, and the fleshy parts are firm. It should not be possible to move the appendages or head easily if enough fluid has been used. An average-sized cat takes about 1 liter. If a syringe is used, it must obviously be refilled. Be sure to cap the base of the needle with your finger when the syringe is taken off, or the fluid will escape. After injecting, remove the needle, and tie the artery.

If embalmed specimens become unusually dry while being dissected they may be moistened with a wetting fluid made as follows:

Phenol (Carbolic acid crystals)	30 g.
Glycerin	250 cc.
Water	1000 cc.

3. INJECTION OF THE CARDIOVASCULAR SYSTEM

The blood vessels of an animal can be rendered more conspicuous by injecting a colored solution into them that will later harden. Colored latex is usually used. If specimens are simply to be preserved, the injection should be made before preservation. But embalmed specimens should be left for several days before injecting the vessels in order to let the embalming solution penetrate the tissues. Injection is made with an injecting syringe and with the needle inserted and tied into a cut in one of the vessels. Avoid injecting air. After injecting, the vessel should be tied off with a piece of string on both sides of the cut.

Inject the arteries of the cat caudally through one of the common carotids until the vessels in the intestine, or those beneath the skin, fill (about 30 cc. for an average-sized adult cat). The systemic veins are injected caudally through the external jugular (30 to 50 cc.), and the hepatic portal system through one of the intestinal veins in the mesentery (12 to 18 cc.). If only the arteries are to be injected, the blood may be allowed to accumulate in the veins, but if the veins are to be injected as well, the blood should be drained off during the arterial injection through a cut in the external jugular vein.

4. INJECTION OF THE LYMPHATIC VESSELS

If the lymphatics of the mammal are to be studied, they can be injected in a freshly killed specimen. A saturated solution of Berlin blue should be injected subcutaneously in various parts of the body, for this substance will enter the lymphatic capillaries, but not those of the cardiovasular system. Injections should be made in all the foot pads, in the pad at the end of the nose, and in the lips. It is necessary to inject slowly and to maintain a pressure on the syringe for 15 to 30 minutes at each site. Also massage the limbs and neck, working toward the center of the body, to aid the flow of the injection fluid. To inject the lymphatics at the base of the mesentery, it is necessary to open the body and to inject into the peripheral parts of the mesenteric lymph nodes. The efferent lymphatic from any lymph node may be injected in the same way.

If a more permanent preparation is desired, a weak solution of warm gelatin may be mixed with the Berlin blue. In this case, the injection must be made while the body and the gelatin are still warm. Injecting with gelatin solution is more difficult.

5. PREPARATION OF SKELETONS

Skeletons should be prepared from mature specimens that have been freshly killed, or from specimens preserved in brine. Fresh specimens are better. First skin and dismember the specimen, removing the head and legs. Then eviscerate the specimen and cut off the larger masses of flesh from the bones. The rest of the flesh can easily be scraped off after it has been loosened in one of several ways. (1) Let it macerate (decay) in a closed container of water at room temperature for several months. (2) Allow insects to remove most of the soft parts. Dermestid beetles are particularly good, and some institutions maintain a colony for this purpose. (3) Simmer the specimen in water, or in a soap solution made by diluting 1 part of the following stock solution with 3 or 4 parts of water:

Ammonia (strong)	150 cc.
Hard soap	75 g.
Potassium nitrate (saltpeter)	12 g.
Water	2000 cc.

The brain can be removed through the foramen magnum with a wire having a flattened loop at one end.

If disarticulated bones are desired, one can let the bones simmer for a considerable time (an hour or two). But if a skeleton in which the bones are held together by ligaments is the objective, one must look at the specimen frequently to be sure that the muscles are soft enough to scrape off, but that the ligaments are not so soft that they detach easily. It is also necessary to cut some of the larger ligaments and tendons, especially those on the underside of the paws of mammals, or they will shorten and distort the specimen.

After the flesh has been removed, it is generally desirable to degrease the bones. Drill a small hole in each end of the larger bones and then place the bones for a day or so in turpentine, benzine, or carbon tetrachloride. Carbon tetrachloride is the most effective, but its fumes are poisonous, so the work should be done under a fume hood. The bones should then be bleached for a day or two in a solution of hydrogen peroxide. When the preparation is cleaned and bleached satisfactorily, the parts of the skeleton must be nailed out in the desired position before the preparation dries.

To get a disarticulated skull, simply continue the boiling process until the sutures are soft enough to permit pulling the bones apart. Another method is to fill the cranial cavity with dried peas, tightly cork the foramen magnum, and put the skull in water. The peas will swell and loosen the bones. Young adults should be used for disarticulated skulls, as some of the skull bones grow together in old specimens.

Appendix III   REFERENCES

The references given below include those cited in the text, the more important studies on the anatomy of the cat and other carnivores (dog, panda), and several key references in comparative and human anatomy. These will serve to introduce the vast literature of vertebrate anatomy. More inclusive bibliographies can be found in standard textbooks and in many of the works cited below.

GENERAL

Alexander, R. M.: Animal Mechanics. London, Sidgwick and Jackson, 1968.

Bolk, L., and others: Handbuch der vergleichenden Anatomie der Wirbeltiere. 6 vols. Berlin and Vienna, Urban and Schwarzenberg, 1931–1938. Reprinted: Amsterdam, A. Asher and Co., 1967.

DeBeer, G. R.: The Vertebrate Skull. Oxford, Clarendon Press, 1937.

Edgeworth, F. H.: The Cranial Muscles of Vertebrates. London, Cambridge University Press, 1935.

Gans, C.: Biomechanics, An Approach to Vertebrate Biology. Philadelphia, J. B. Lippincott Company, 1974.

Goodrich, E. S.: Studies on the Structure and Development of Vertebrates. London, Macmillan and Co., 1930. Reprinted: New York, Dover Publications, 1958.

Grassé, P.-P., (ed.): Traité de Zoologie. Vols. 11-17 deal with protochordates and vertebrates. Paris, Masson et Cie., 1948–1958.

Hildebrand, M.: Anatomical Preparations. Berkeley, University of California Press, 1969.

Hildebrand, M.: Analysis of Vertebrate Structure. New York, John Wiley and Sons, 1974.

International Committee on Veterinary Anatomical Nomenclature: Nomina Anatomica Veterinaria. 2nd ed. Vienna, Adolf Holzhausen's Successors, 1972. Distributed in U.S.A. by Dept. of Anatomy, New York State Veterinary College, Ithaca.

Jaeger, E. C.: A Source-Book of Biological Names and Terms. 3rd ed. Springfield, Ill., Charles C Thomas, 1955.

Kappers, C. U. A., Huber, G. C., and Crosby, E. C.: The Comparative Anatomy of the Nervous System of Vertebrates, Including Man. 2 vols., New York, The Macmillan Company, 1936.

Kopsch, K., and Knese, K. H.: Nomina Anatomica. Vergleichenden übersicht der Basler, Jenaer und Pariser Nomenklatur. Stuttgart, Georg Thieme Verlag, 1957.

Romer, A. S.: The Vertebrate Body. 4th ed. Philadelphia, W. B. Saunders Company, 1970.

Romer, A. S.: The Vertebrate Story. Chicago, University of Chicago Press, 1958.

Schmidt-Nielsen, K.: How Animals Work. Cambridge, Cambridge University Press, 1972.

Walker, W. F., Jr.: Vertebrate Dissection. 5th ed. Philadelphia, W. B. Saunders Company, 1975.

Walls, G. L.: The Vertebrate Eye and Its Adaptive Radiation. Bloomfield Hills, Cranbrook Institute of Science, Bull. No. 19, 1942.

Webster, D., and Webster, M.: Comparative Vertebrate Morphology. New York, Academic Press, 1974.

Wessells, N. K. (ed.): Vertebrate Adaptations. San Francisco, W. H. Freeman and Company, 1969.

Wilder, B. G., and Gage, S. H.: Anatomical Technology as Applied to the Domestic Cat. New York, A. S. Barnes and Company, 1882.

Young, J. Z.: The Life of Vertebrates. 2nd ed. London, Oxford University Press, 1962.

CAT, CARNIVORES, AND HUMAN BEINGS

Abell, N. B.: A comparative study of the variations of the postrenal vena cava of the cat and rat and a description of two new variations. Denison University Journal of the Science Laboratory, 1947, vol. 40, pp. 87–117.

Anthony, C. P., and Kolthoff, N. J.: Textbook of Anatomy and Physiology. 8th ed. St. Louis, The C. V. Mosby Co., 1971.

Arey, L. B.: Developmental Anatomy. 7th ed. Philadelphia, W. B. Saunders Company, 1974.

Barry, A.: The aortic arch derivatives in the human adult. Anatomical Record, vol. III, pp. 221–238, 1951.

Bloom, W., and Fawcett, D. W.: A Textbook of Histology. 10th ed. Philadelphia, W. B. Saunders Company, 1975.

Crafts, R. C.: A Textbook of Human Anatomy. New York, The Ronald Press Company, 1966.

Crouch, J. E.: Functional Human Anatomy. Philadelphia, Lea and Febiger, 1965.

Crouch, J. E.: Text-Atlas of Cat Anatomy. Philadelphia, Lea & Febiger, 1969.

Davis, D. D.: The giant panda, a morphological study of evolutionary mechanisms. Fieldiana, Zoological Memoirs, vol. 3, pp. 1–340, 1964.

Davis, D. D., and Story, H. E.: The carotid circulation in the domestic cat. Zoological Series Field Museum of Natural History, vol. 28, pp. 1–47, 1943.

Dienhart, C. H.: Basic Human Anatomy and Physiology. 2nd ed. Philadelphia, W. B. Saunders Company, 1973.

Field, H. E., and Taylor, M. E.: An Atlas of Cat Anatomy. Chicago, Chicago University Press, 1954.

Gardner, W. D., and Osburn, W. A.: Structure of the Human Body. 2nd ed. Philadelphia, W. B. Saunders Co., 1973.

Gilbert, S. G.: Pictorial Anatomy of the Cat. Seattle, University of Washington Press, 1968.

Huntington, G. S., and McClure, C. F. W.: The development of the veins in the domestic cat. Anatomical Record, vol. 20, pp. 1–31, 1920.

Jacob, S. W., and Francone, C. A.: Structure and Function in Man. 3rd ed. Philadelphia, W. B. Saunders Company. 1974.

Jayne, H.: Mammalian Anatomy. Part I. The Skeleton of the Cat. Philadelphia, J. B. Lippincott Company, 1898.

Kerr, N. S.: The homologies and nomenclature of the thigh muscles of the opossum, cat, rabbit, and rhesus monkey. Anatomical Record, vol. 121, pp. 481–493, 1955.

King. B. G., and Showers, M. J.: Human Anatomy and Physiology. 6th ed. Philadelphia, W. B. Saunders Company, 1969.

Leach, W. J.: Functional Anatomy, Mammalian and Comparative. 3rd ed. New York, McGraw-Hill Book Company, 1961.

Miller, M. E., Christensen, G. C., and Evans, H. E.: Anatomy of the Dog. Philadelphia, W. B. Saunders Company, 1964.

Moore, K. L.: The Developing Human. Philadelphia, W. B. Saunders Company, 1973.

Northcutt, R. G., Kenneth, L. W., and Barber, R. P.: Atlas of the Sheep Brain. 2nd ed. Champaign, Illinois, Stiles Publishing Company, 1966.

Ranson, S. W.: The Anatomy of the Nervous System. 10th ed. Revised by Clark, S. L. Philadelphia, W. B. Saunders Company, 1959.

Rasmussen, A. T.: The Principal Nervous Pathways. 4th ed. New York, The Macmillan Company, 1952.

Reighard, J. E., and Jennings, H. S.: Anatomy of the Cat. 3rd ed. Revised by Elliott, R. New York, Henry Holt and Company, Inc., 1935.

Sisson, S., and Grossman, J. H.: The Anatomy of the Domestic Animals. 5th ed. Revised by Getty, R. Philadelphia, W. B. Saunders Company, 1975.

Warwick, R., and Williams, R. L. (eds): Gray's Anatomy. 35th British ed. Philadelphia, W. B. Saunders Co., 1973.

Yoshikawa, T.: Atlas of the Brains of Domestic Animals. University Park, Pennsylvania State University Press, 1968.

Young, J. Z., assisted by Hobbs, M. J.: The Life of Mammals. 2nd ed. New York, Oxford University Press, 1975.

INDEX

Many terms are indexed under their determining noun: e.g., artery, bone, foramen, gland, muscle, vein, and so forth. Structures in figures are indexed only when they appear in figures not referred to during the discussion of the structure.